American Book Company
Meeting Standards,
Exceeding Expectations

MW00677149

Dear Educator,

Thank you for your interest in American Book Company's state-specific test preparation resources. We commend you for your interest in pursuing your students' success. Feel free to contact us with any questions about our books, software, or the ordering process.

Our Products Feature	Your Students Will Improve
Multiple-choice and open-ended diagnostic tests	Confidence and mastery of subjects
Step-by-step instruction	Concept development
Frequent practice exercises	Critical thinking
Chapter reviews	Test-taking skills
Multiple-choice practice tests	Problem-solving skills

American Book Company's writers and curriculum specialists have over 100 years of combined teaching experience, working with students from kindergarten through middle, high school, and adult education.

Our company specializes in effective test preparation books and software for high stakes graduation and grade promotion exams across the country.

How to Use This Book

Each book:

*contains a chart of standards which correlates all test questions and chapters to the state exam's standards and benchmarks as published by the state department of education. This chart is found in the front of all preview copies and in the front of all answer keys.

*begins with a full-length pretest (diagnostic test). This test not only adheres to your specific state standards, but also mirrors your state exam in weights and measures to help you assess each individual student's strengths and weaknesses.

*offers an evaluation chart. Depending on which questions the students miss, this chart points to which chapters individual students or the entire class need to review to be prepared for the exam.

*provides comprehensive review of all tested standards within the chapters. Each chapter includes engaging instruction, practice exercises, and chapter reviews to assess students' progress.

*finishes with two full-length practice tests for students to get comfortable with the exam and to assess their progress and mastery of the tested standards and benchmarks.

While we cannot guarantee success, our products are designed to provide students with the concept and skill development they need for the graduation test or grade promotion exam in their own state. We look forward to hearing from you soon.

Sincerely,

The American Book Company Team

PO Box 2638 ★ Woodstock, GA 30188-1383 ★ Phone: 1-888-264-5877 ★ Fax: 1-866-827-3240

Georgia 4th Grade CRCT Test in Science
Chart of Standards

Mastering the Georgia 4th Grade CRCT Test in Science

Chart of Standards

The following chart correlates each question on the Diagnostic Test, Practice Test 1, and Practice Test 2 to the Science **GPS Standards published by the Georgia Department of Education**. These test questions are also correlated with chapters in the Mastering the 4th Grade CRCT Test in Science.

Competency Standards Earth Science	Chapter Number	Diagnostic Test Questions	Practice Test 1 Questions	Practice Test 2 Questions
S4E1: Students will compare and contrast the physical attributes of stars, star patterns and planets.				
a. Recognize the physical attributes of stars in the night sky such as number, size, color and patterns.	1	4, 44, 67	30, 40	8, 27
b. Compare the similarities and differences of planets to the stars in appearance, position, and number in the night sky.	2	30	32	
c. Explain why the pattern of stars in a constellation stays the same, but a planet can be seen in different locations at different times.	1		22, 55	13
d. Identify how technology is used to observe distant objects in the sky.	2	14, 51	3, 14	10, 38

Chart of Standards

Competency Standards	Chapter Number	Diagnostic Test Questions	Practice Test 1 Questions	Practice Test 2 Questions
S4E2: Students will model the position and motion of the Earth in the solar system and will explain the role of relative position and motion in determining sequence of the phases of the Moon.				
a. Explain the day/night cycle of the Earth using a model	4	11	36	17, 50
b. Explain the sequence of the phases of the Moon.	3	18, 54	25, 48	35, 62
c. Demonstrate the revolution of the Earth around the Sun and the Earth's tilt to explain the seasonal changes.	4	39, 64	38, 53	24, 57, 69
d. Demonstrate the relative size and order from the Sun of the planets in the solar system.	3	10, 41	5	19, 65, 68, 69
S4E3: Students will differentiate between the states of water and how they related to the water cycle and weather.				
a. Demonstrate how water changes states from solid (ice) to liquid (water) to gas (water vapor/steam) and changes from gas to liquid to solid.	5, 7	16, 47	17, 43	44
b. Identify the temperatures at which water becomes a solid and at which water becomes a gas.	5		58, 66	40
c. Investigate how clouds are formed.	6	1, 21, 22	41	5, 22

Competency Standards	Chapter Number	Diagnostic Test Questions	Practice Test 1 Questions	Practice Test 2 Questions
d. Explain the water cycle (evaporation, condensation and precipitation)	7	53		
e. Investigate different forms of precipitation and sky conditions. (rain, snow, sleet, hail, clouds and fog).	6	22, 26, 62	34, 51	22, 33
S4E4: Students will analyze weather charts/maps and collect weather data to predict weather events and infer patterns and seasonal changes.				
a. Identify weather instruments and explain how each is used in gathering weather data and making forecasts (thermometer, rain gauge, barometer, wind vane, anemometer).	9	6, 31, 49	10, 11, 62	1, 2, 29, 31, 52
b. Using a weather map, identify the fronts, temperature and precipitation and use the information to interpret the weather conditions.	8	33, 36, 69	28, 45, 46	14, 42, 46
c. Use observations and records of weather conditions to predict weather patterns throughout the year.	9			
d. Differentiate between weather and climate.	8	56	20	

Chart of Standards

Competency Standards Physical Science	Chapter Number	Diagnostic Test Questions	Practice Test 1 Questions	Practice Test 2 Questions
S4P1: Students will investigate the nature of light using tools such as mirrors, lenses and prisms.				
a. Identify materials that are transparent, opaque and translucent.	10	25, 42, 46, 52, 57	27, 33, 39, 57	11, 26, 51, 60
b. Investigate the reflection of light using a mirror and a light source.	10	65	19, 42	36
c. Identify the physical attributes of a convex lens, a concave lens and a prism and where each is used.	11	19	31	34
S4P2: Students will demonstrate how sound is produced by vibrating objects and how sound can be varied by changing the rate of vibration.				
a. Investigate how sound is produced.	12	3, 38, 61	13, 23, 52	4, 6, 41, 63
b. Recognize the conditions that cause pitch to vary.	12	9, 23, 32, 59	6, 49, 65, 68	16, 23, 55
S4P3: Students will demonstrate the relationship between the application of a force and the resulting change in position and motion of an object.				
a. Identify simple machines and explain their uses (lever, pulley, wedge, incline plane, screw, wheel and axel).	13	13, 48, 68	1, 61	332, 43
b. Using different size objects, observe how force affects speed and motion.	14	5, 28, 70	60, 63, 70	48, 58, 70
c. Explain what happens to the speed or direction of an object when a greater force than the initial one is applied.	14	5, 70	69, 70	48, 70

Competency Standards Physical Science	Chapter Number	Diagnostic Test Questions	Practice Test 1 Questions	Practice Test 2 Questions
d. Demonstrate the effect of derivational force on the motion of an object.	14	34	8, 15	20, 25

Chart of Standards

Competency Standards Life Science	Chapter Number	Diagnostic Test Questions	Practice Test 1 Questions	Practice Test 2 Questions
S4L1: Students will describe the roles of organisms and the flow of energy within an ecosystem.				
a. Identify the roles of producers, consumers and decomposers in a community.	15	2, 8, 15, 37, 63	4, 9, 59, 67	7, 9, 30, 37, 56
b. Demonstrate the flow of energy throughout a food web/food chain beginning with sunlight and including producers, consumers and decomposers.	16	24, 29, 40, 55, 58, 66	2, 12, 21, 29, 50, 64	3, 12, 15, 18, 21, 47
c. Predict how changes in the environment would affect a community (ecosystem) of organisms.	17, 18	12, 20, 50, 60	24, 44, 47, 54	39, 53, 61, 64
d. Predict effects on a population if some of the plants and animals in the community are scarce or if there are too many.	18	7, 35, 45	35, 37, 56	28, 45, 54, 67
S4L2: Students will identify factors that affect the survival or extinction of organisms such as adaptation, variation of behaviors (hibernation), and external features (camouflage and protections).				
a. Identify external features of organisms that allow them to survive or reproduce better than organisms that do not have these features (for example: camouflage, use of hibernation, protection, etc.).	17	17, 27, 43	16, 18, 26	59, 66
b. Identify factors that may have led to the extinction of some organisms.	18			

American Book Company's

MASTERING THE

GEORGIA 4TH GRADE CRCT

IN SCIENCE

Written to GPS 2006 Standards

Liz Thompson
Michelle Gunter
Emily Powell

American Book Company
PO Box 2638
Woodstock, GA 30188-1383
Toll Free: 1 (888) 264-5877   Phone: (770) 928-2834
Fax: (770) 928-7483 Toll Free Fax: 1 (866) 827-3240
Web site: www.americanbookcompany.com

ACKNOWLEDGEMENTS

The authors would like to gratefully acknowledge the formatting and technical contributions of Becky Wright.

We also want to thank Mary Stoddard and Eric Field for their expertise in developing the graphics for this book.

A special thanks to Marsha Torrens for her editing assistance.

A final thank you goes to Dr. Karen Michael for her review of this text. Dr. Michael is an Associate Professor of Early Childhood Education at Mercer University in the Tift College of Education.

Printed in the United States of America

08/08

Table of Contents

Using a Science Journal

A science journal is an important tool to use. It is a bound notebook where you write your experiences. Science journals work best when they are used each day. In this way, it records your science experiences.

Sometimes a simple writing about the lesson is all that is needed. Other times you complete a more formal activity.

The following steps tell you how to make your science journal:

1   **Leave the first few pages blank. They will become a table of contents.**

2   **Put the date and objective on the left hand side of the journal.**

3   **Below the date, write important notes about the procedure or topic.**

4   **Put your data or other notes/drawings on the right hand side of the journal.**

5   **Below the data, write a brief summary of the activity. The summary explains what you found.**

Your science journal should look like the drawing on the following page.

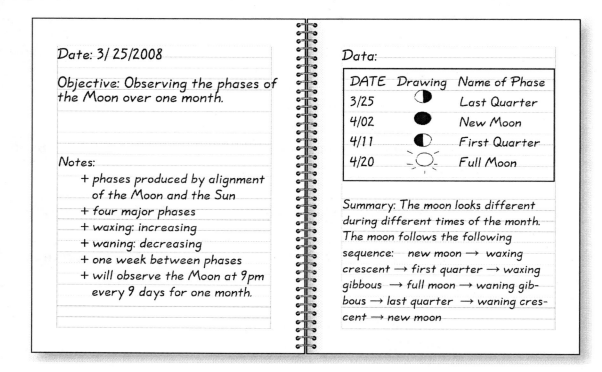

Date: 3/ 25/2008

Objective: Observing the phases of the Moon over one month.

Notes:
+ phases produced by alignment of the Moon and the Sun
+ four major phases
+ waxing: increasing
+ waning: decreasing
+ one week between phases
+ will observe the Moon at 9pm every 9 days for one month.

Data:

DATE	Drawing	Name of Phase
3/25		Last Quarter
4/02		New Moon
4/11		First Quarter
4/20		Full Moon

Summary: The moon looks different during different times of the month. The moon follows the following sequence: new moon → waxing crescent → first quarter → waxing gibbous → full moon → waning gibbous → last quarter → waning crescent → new moon

Preface

Mastering the Georgia 4th Grade CRCT in Science will help students who are learning or reviewing material for the Georgia test that is now required for each gateway or benchmark course. **The materials in this book are based on the Georgia Performance Standards as published by the Georgia Department of Education.**

This book contains several sections. These sections are as follows: 1) General information about the book; 2) A Diagnostic Test and Evaluation Chart; 3) Domains/Chapters that teach the concepts and skills to improve readiness for the Georgia 4th Grade CRCT Test in Science; 4) Two Post/Practice Tests. Answers to the tests and exercises are in a separate manual. The answer manual also contains a Chart of Standards for teachers to make a more precise diagnosis of student needs and assignments.

We welcome comments and suggestions about the book. Please contact us at

American Book Company
PO Box 2638
Woodstock, GA 30188-1383

Toll Free: 1 (888) 264-5877
Phone: (770) 928-2834
Fax: (770) 928-7483
Web site: www.americanbookcompany.com

About the Authors

Liz A. Thompson holds a B.S. in Chemistry and an M.S. in Analytical Chemistry, both from the Georgia Institute of Technology. Research conducted as both an undergraduate and a graduate student focused on the creation and fabrication of sensors based on conducting polymers and biomolecules. Post graduate experience includes work in radioanalytical chemistry. Her publications include several articles in respected scientific journals, numerous test preparation manuals for middle and high school science students, as well as partial authorship of the graduate level textbook *Radioanalytical Chemistry* (2007). She is also editor of the upcoming text *Sensors* (in press). At every educational level, Mrs. Thompson has enjoyed teaching, tutoring and mentoring students in the study of science.

Michelle Gunter graduated from Kennesaw State University in Kennesaw, Georgia with a B.S. in Secondary Biology Education. She is a certified teacher in the field of Biology in the state of Georgia. She has three years experience in high school science classrooms. She has nine years experience in biology and biological systems. She has won awards for her research in the field of aquatic toxicology. Mrs. Gunter enjoys teaching students of all ages the wonders of the natural world. Mrs. Gunter is currently pursuing her M.S. in Biology at Georgia State University.

Emily J. Powell holds a B.S. and an M.S. in Physical Geography with an emphasis in atmospheric science, both from the University of Georgia. Her graduate research focused on the surface mass balance of the Greenland Ice Sheet, comparing ERS satellite and model data. She holds partial authorship of a published research paper in a respected meteorological journal. She has five years experience in meteorology product development. She has two years experience teaching and tutoring students of all ages in classroom and professional environments in both the United States and Spain.

Georgia 4th Grade CRCT Diagnostic Test in Science

1 What type of cloud forms at the highest level in the atmosphere?   S4E3c

A   cumulus
B   nimbus
C   cirrus
D   stratus

2 What is a producer's role in a community?   S4L1a

A   to eat food
B   to eat wastes
C   to make food
D   to remain camouflaged

3 Where will sound travel the fastest?   S4P2a

A   under the ocean
B   through the air
C   in space
D   through a rock mountain

4 How does the color of a star change?   S4E1a

A   in relation to its distance from Earth
B   in relation to its distance from the Sun
C   in relation to the sound waves it receives
D   in relation to its size and temperature

5 What does NOT directly affect the motion of an object?   S4P3b, c

A   direction of force applied
B   strength of force applied
C   mass of the object
D   heat of the object

6 **What measurement tool is this?**

S4E4a

A   thermometer   C   anemometer
B   barometer   D   rain gauge

7 **How will an overabundance of caribou affect the animals pictured here?**

S4L1d

A   Its population will increase.
B   Its population will decrease.
C   Its population will stay the same.
D   Its population will become scarce.

8 **A pine tree uses its needles to collect sunlight for energy. What type of organism is a pine tree?**

S4L1a

A   a producer   C   a decomposer
B   a consumer   D   a predator

9 **Which point on the violin string will make the highest pitched sound?** S4P2b

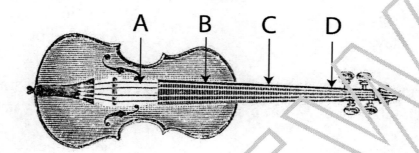

A Point A	C Point C
B Point B	D Point D

10 **What is the largest planet in our solar system?** S4E2d

A Jupiter	C the Sun
B Earth	D an asteroid

11 **Sunrise and sunset are caused by** S4E2a

A the revolution of the Earth.
B the revolution of the Sun.
C the rotation of the Earth.
D the rotation of the Sun.

12 **Both goats and giant Galapagos Tortoises eat plants. During the 1800s, sailors released goats on the Galapagos Islands. The release of the goats created a scarcity of which organisms?** S4L1c

A plants only
B tortoises only
C both plants and tortoises
D the goats did not create any scarcities

13 Mark used a screwdriver to open a can of paint. What type of simple machine did he use?

S4P3a

A   screw

C   wedge

B   lever

D   inclined plane

14 Jeral is observing the difference between planets and stars for a school project. What instrument is he MOST likely using?

S4E1

A   binoculars

C   telescope

B   space station

D   probe

15 Which organism below is a producer?

S4L1a

A

C

B

D

16 What part of the water cycle dries water from the ground?

S4E3a

A   precipitation

C   condensation

B   evaporation

D   circulation

17 What type of adaptation is a turtle's shell?   S4L2a

A   protective   C   hibernation
B   mimicry   D   warning coloration

18 When is the Moon the LEAST visible?   S4E2b

A   at full moon   C   at third quarter
B   at half moon   D   at new moon

19 What type of lens is thick in the middle?   S4P1c

A   convex   C   prism
B   concave   D   kaleidoscope

20 Which organism is well adapted to an abundance of sunlight, little rain and high temperatures?   S4L1c

A   polar bear   C   penguin
B   red swamp lily   D   cactus

21 What happens when liquid water droplets in a cloud grow bigger and heavier?   S4E3c

A   They fall from the cloud as dew.
B   They fall from the cloud as rain.
C   They evaporate out of the cloud.
D   The cloud sinks towards the ground and becomes fog.

22 Which of the following is NOT needed to form a cloud?   S4E3c,e

A   small particles in the air   C   rising air
B   water vapor   D   high pressure

23 **Brass instruments make sounds from the vibrations of the player's lips. What feature of a brass instrument will determine the pitch of the sound?**

S4P2b

   A   the length of the brass tubing
   B   the color of the brass tubing
   C   the mass of the brass tubing
   D   the weight of the brass tubing

24 **What type of consumers are humans?**

S4L1b

   A   herbivore   C   omnivore
   B   carnivore   D   decomposer

25 **Which object is NOT transparent?**

S4P1a

   A   glass   C   plastic wrap
   B   diamond   D   cardboard

26 **What is liquid water falling from the sky called?**

S4E3c

   A   snow   C   hail
   B   sleet   D   rain

27 **What type of adaptation below would MOST help an animal survive in extreme cold?**

S4L2a

   A   thin fur   C   thick fat layer
   B   fast movement   D   sharp teeth

28 **A medicine ball is a large massive ball. A beach ball is a large light ball. If Thomas throws the medicine ball and the beach ball with the same force which will travel farther?**

S4P3b

   A   the medicine ball will travel farther
   B   the beach ball will travel farther
   C   both balls will travel the same distance
   D   the massive object will travel farther

29 Which organism in the food web pictured here gets its energy directly from the Sun?

S4L1b

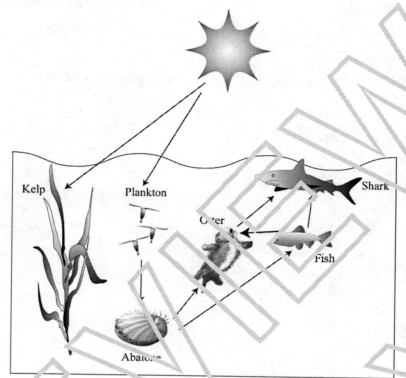

A   fish

B   kelp

C   otter

D   shark

30 Which of the following BEST describes the movement of stars?

S4E1b

A   Stars rotate around the Earth.

B   Stars always change their positions.

C   Stars revolve around the Earth.

D   Stars remain in the same positions.

31 A storm system to the west is moving east toward Georgia. What instrument would be MOST useful to monitor the approach of the storm?

S4E4a

A   a rain gauge

B   an anemometer

C   a thermometer

D   a barometer

32 Which noun BEST describes the pitch of a sound?

S4P2b

A frequency

B loudness

C pace

D volume

33 Which weather map shows a correct placement of a cold front?

S4P4b

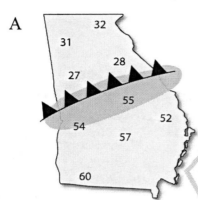

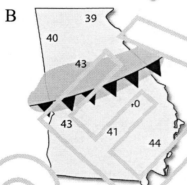

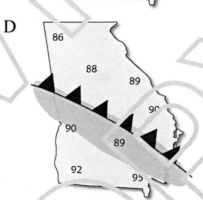

34 Hermione is on top of the Empire State Building. She drops a penny. What force causes the penny to fall to the street below?

S4P3d

A the force she threw it with

B the force of gravity

C the force of wind

D the force of sunlight

35 How will a scarcity of plants affect the animal pictured here?

S4L1d

Donkey

A   Its population will increase.

B   Its population will decrease.

C   Its population will stay the same.

D   Its population will become abundant.

36 Pepe walked to school one morning. It was very warm and cloudy. While Pepe was in school, high pressure moved in, with a cold front. When Pepe walked home, what was the weather probably like?

S4E4c

A   warmer and wet

B   cooler and wet

C   cooler and stormy

D   cooler and dry

37 What do mice do in a community?

S4L1a

A   make food

B   consume food

C   break down wastes

D   hide other organisms

38 Ronan is trying to whisper or speak softly. How can he accomplish this?   S4P2a

    A   by making his vocal cords shorter

    B   by making his vocal cords longer

    C   by forcing more air through his vocal cords

    D   by forcing less air through his vocal cords

39 During summer in the Northern Hemisphere, the Sun's rays are MOST concentrated on the   S4E2c

    A   poles.   C   Northern Hemisphere.

    B   equator.   D   Southern Hemisphere.

40 Snakes always kill and eat other animals. What type of consumer is a snake?   S4L1b

    A   herbivore   C   omnivore

    B   carnivore   D   decomposer

41 Which of the following correctly lists the "Gas Giants" in order from closest to the Sun to farthest from the Sun?   S4E2d

    A   Saturn, Uranus, Neptune, Jupiter

    B   Jupiter, Uranus, Saturn, Neptune

    C   Jupiter, Saturn, Uranus, Neptune

    D   Uranus, Jupiter, Saturn, Neptune

42 Which color of white light contains the LEAST energy?   S4P1a

    A   red   C   green

    B   black   D   orange

43 When winter comes to the arctic, temperatures can drop below −40°C. How do polar bears adapt to this change?   S4L2a

    A   They camouflage.   C   They get smaller.

    B   They hibernate.   D   They die.

44 Stars are classified according to their S4E1a

    A   shape and distance.   C   color and distance.
    B   color and size.   D   size and rotation.

45 A scarcity of which organism below would MOST harm a population of rattlesnake? S4L1d

    A

    C

    B

    D

46 Which pair of sunglasses will absorb the MOST light? S4P1a

    A   a pair of black sunglasses
    B   a pair of red sunglasses
    C   a pair of yellow sunglasses
    D   a pair of pink sunglasses

47 In which situation would evaporation MOST likely occur? S4E3a

    A   A bag of ice left in a cooler.
    B   A cup of water left on a sunny picnic table.
    C   A cup of water left in an air conditioned room.
    D   A sealed soda bottle left in an air conditioned room.

48 What simple machine is shown in the picture here?

S4P3a

A   screw   C   wheel and axel
B   lever   D   inclined plane

49 A wind is moving from the south to north. What direction is your weather vane pointing?

S4J4a

A   south   C   east
B   north   D   west

50 Brown tree snakes were brought to an island by mistake. They eat birds, eggs, frogs and mice. They can easily climb trees and live in the canopy of MOST forests. Which bird population would be MOST negatively impacted by the brown tree snake?

S4L1c

A   Birds that nest on the ground.
B   Birds that nest in trees.
C   Birds that nest on cliffs.
D   Birds that nest at the beach.

51 An instrument that collects light from objects in space is called a(n)

S4E1d

A   anemometer.   C   barometer.
B   probe.   D   telescope.

52 Why do white objects stay the coolest when placed in sunlight?

   A   They absorb more light energy.

   B   They reflect less light energy.

   C   They reflect more light energy.

   D   They refract less light energy.

53 What causes water vapor in the air to become visible as clouds?

   A   evaporation   C   respiration

   B   transpiration   D   condensation

54 Which of the following correctly lists the sequence of the phases of the Moon?

   A   new moon, first quarter, full moon, last quarter

   B   first quarter, new moon, third quarter, full moon

   C   first quarter, half moon, third quarter, full moon

   D   new moon, first quarter, third quarter, full moon

55 Which organism below is NOT prey?

   A   grass   C   rabbit

   B   moth   D   snake

56 Esmeralda records the temperature, cloud cover, and precipitation in her journal every day. What is she observing?

   A   the weather

   B   the climate

   C   the constellations

   D   the water cycle

57 It is wintertime. You are going skiing outside. You want to stay warm. What color shirt should you wear?

S4P1a

A   black

B   white

C   red

D   blue

58 Algae live in a pond. It uses sunlight to make food. What type of organism is algae?

S4L1c

A   a producer

B   a consumer

C   a decomposer

D   a predator

59 How can you make a low pitched sound with your vocal cords?

S4P2b

A   make them shorter

B   make them longer

C   force more air through them

D   force less air through them

60 Which animal below would be LEAST impacted by a cold, wet winter?

S4L1c

A   tortoise

B   jackrabbit

C   elephant

D   penguin

61 Which type of matter transmits sound the fastest?

S4P2a

A   gas

B   liquid

C   solid

D   vacuum

62 One type of frozen precipitation falls as rain but freezes before it hits the ground. What is it called?

S4E3e

A   snow

B   hail

C   fog

D   sleet

63 What does a termite do in a community? S4L1a

    A   makes food   C   breaks down wastes

    B   consumes food   D   hides other organisms

64 Which of the following does NOT contribute to seasonal changes on Earth? S4E2c

    A   the tilt of the Earth's axis

    B   the revolution of the Earth around the Sun

    C   the sequence of the phases of the Moon

    D   the relationship between the Earth and the Sun

65 Which will refract light the LEAST? S4P1b

    A   air   C   glass prism

    B   water   D   lump of coal

66 How would you BEST summarize the difference between a food chain and a food web? S4L1b

    A   A food chain is a complicated food web.

    B   A food web represents several food chains.

    C   A food chain only shows small animals.

    D   A food web shows only animals in a community.

67 The biggest stars are also S4E1a

    A   the hottest stars.   C   the coolest stars.

    B   the dullest stars.   D   the reddest stars.

68 A doorknob is an example of what type of simple machine? S4P3a

    A   screw   C   wheel and axel

    B   pulley   D   inclined plane

69 **What type of weather is typical when a warm front passes and low pressure moves in?**

S4E4b

A   warm and rainy   C   cold and rainy

B   warm and dry   D   cold and dry

70 **What can forces NOT do to an object?**

S4P3b, c

A   cause it to start moving

B   cause it to stop moving

C   cause it to gain mass

D   cause it to change direction

Domain One
Earth Science

Chapter 1: The Night Sky

S4E1a: Recognize the physical attributes of stars in the night sky such as number, size, color and patterns.

S4E1c: Explain why the pattern of stars in a constellation stays the same, but a planet can be seen in different locations at different times.

Chapter 2: Observing Planets and Stars

S4E1b: Compare the similarities and differences of planets to the stars in appearance, position and number in the night sky.

S4E1d: Identify how technology is used to observe distant objects in the sky.

Chapter 3: The Solar System and Moon

S4E2b: Explain the sequence of the phases of the Moon.

S4E2d: Demonstrate the relative size and order from the Sun of the planets in the solar system.

Chapter 4: Rotation and Revolution

S4E2a: Explain the day/night cycle of the Earth using a model

S4E2c: Demonstrate the revolution of the Earth around the Sun and the Earth's tilt to explain the seasonal changes.

Chapter 5: States of Water

S4E3a: Demonstrate how water changes states from solid (ice) to liquid (water) to gas (water vapor/steam) and changes from gas to liquid to solid.

S4E3b: Identify the temperatures at which water becomes a solid and at which water becomes a gas.

Chapter 6: Clouds

S4E3c: Investigate how clouds are formed.

S4E3e: Investigate different forms of precipitation and sky conditions. (rain, snow, sleet, hail, clouds and fog).

Chapter 7: The Water Cycle

S4E3a: Demonstrate how water changes states from solid (ice) to liquid (water) to gas (water vapor/steam) and changes from gas to liquid to solid.

S4E3d: Explain the water cycle (evaporation, condensation and precipitation).

Chapter 8: Weather Patterns

S4E4b: Using a weather map, identify the fronts, temperature and precipitation and use the information to interpret the weather conditions.

S4E4d: Differentiate between weather and climate.

Chapter 9: Instruments in Weather Forecasting

S4E4a: Identify weather instruments and explain how each is used in gathering weather data and making forecasts (thermometer, rain gauge, barometer, wind vane, anemometer).

S4E4c: Use observations and records of weather conditions to predict weather patterns throughout the year.

Chapter 1
The Night Sky

THE STARS

The **universe** is the space that contains every piece of matter and every bit of energy that has ever existed. It includes everything you see around you, everything on Earth, all the planets nearby and all the stars in the sky. The universe is really big, but it is limited in size. It is round in shape. It is bigger than we can see or imagine. The universe is a pretty big place!

The universe is filled with stars. **Stars** are huge balls of hot gas. Stars give off energy in the form of light and heat. We see their energy as bright white points in the night sky. You may notice that all of the points of light are not the same. Some are bigger than others. Some are bright and some are dim. In fact, stars are not all white. They can appear in a range of colors. Examine Figure 1.1. It tells us some information about stars and their color.

Class	Apparent Color
O	blue
B	blue
A	blue-white
F	white
G	yellow-white
K	orange
M	red

HOTTEST

COLDEST

Figure 1.1 Classifying Stars

CLASSIFYING STARS

Stars are very, very far away. The closest star to Earth is the Sun. Our Sun is a class G star (see figure 1.2 below.) Figure 1.1 tells us what you have probably already noticed: our Sun looks yellow-white in color.

Why do different stars have different colors? Well, stars do not live forever. They have a lifetime. They are born. They live and then they die. During this lifetime, the stars change.

Stars are formed from clouds of gas in space (mostly hydrogen). After they form, they spend most of their lifetime as **main sequence** stars. You can think of this as an adult star. Main sequence stars are classified by their color and size. Larger stars have more gases. They have more material to burn and are hotter. This also makes them brighter. Smaller stars are cooler and less bright. So, the color of a star has to do with its temperature and size.

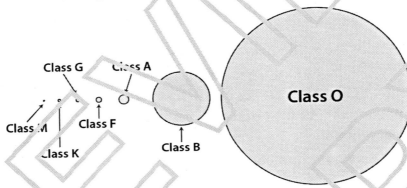

Figure 1.2 Comparison of the Size and Color of Stars

As stars get old, they run out of material to burn. Less material means less heat. They begin to cool. As they cool, their brightness fades.

THE GALAXY

Stars are very big and bright. We can see them from very far away. Most of the stars visible in the night sky are located in other galaxies!

A **galaxy** is a huge system of stars. The shape of a galaxy results from the location of the stars. Stars are not evenly spaced, so the shape of a galaxy can be a little strange.

Our galaxy is the **Milky Way Galaxy**. It is a flattened disc of BILLIONS of stars. It's hard to imagine just how many this is. It is more stars than people could count one-at-a-time in their lifetime! Our Sun is just one of those stars. It is located on an outer arm of the galaxy.

Figure 1.3 Milky Way Galaxy

MILKY WAY GALAXY

We can see our own galaxy from Earth. It looks like a milky band of stars in the night sky. This band divides the night sky in half. All of the other stars that we see are from other galaxies very far away.

Astronomers measure the distance between things in space in light-years. A **light-year** is the distance light travels in one year. They do not measure the distance between things in space in miles or kilometers. The numbers would be HUGE. Just one light-year equals 9.2 billion kilometers!

We don't feel heat from other stars because they are so far away. We can look at them because of the light they give off. We can study them by studying changes in their light.

The Sun is much closer, so we feel its heat and see its light. The Sun appears bigger and brighter than all other stars because it is much closer to Earth. The distance from the Sun to Earth is much less than a light-year. We measure it in kilometers (km). The average distance between the Earth and the Sun is 149.6 million km.

CONSTELLATIONS

The stars are very far away. They appear to stay in the same positions. However, each night, the entire sky can be seen rotating around the North Star. The **North Star**, also called Polaris, is almost directly overhead if you are standing anywhere in the Northern Hemisphere. It marks true north. People in the Southern Hemisphere cannot see the North Star.

We built our first map of the stars using just our eyes. These were the constellations. A **constellation** is a grouping of stars in a region of the sky. One you might know is Orion. **Orion** is sometimes called the Hunter. Orion is not really a hunter in the sky. It is a group of stars that makes a picture. The pictures contained in constellations are made up by humans. These pictures or maps link stars together. They help people remember an area of the sky.

The position of stars relative to each other stays the same. That is, the shapes of the constellations do not change. Early sailors, like Christopher Columbus, used the constellations to find their way around the globe. Figure 1.4 shows the constellation Orion.

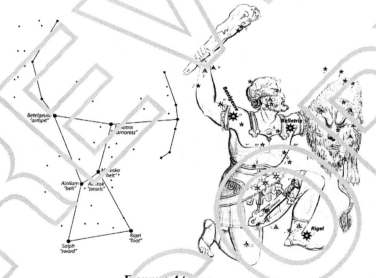

Figure 1.4 Orion

As the entire sky moves each night, constellations move with it. That means the night sky looks very different from day to day and month to month.

Try looking up at a constellation in the sky in the early evening. Then look for it again a few hours later. You will see that it appears to have moved. Really, the stars do not move; it is the Earth that moved. Stars move slowly as the galaxy rotates. They are so far from Earth that their movement is not visible.

Constellations can be seen for a few months at a time. They move in a predictable pattern across the sky. Planets are not part of the constellations, and their movements are not as predictable. The movement of a planet looks different from the movement of constellations. Keep in mind, the stars are not really moving, the Earth is!

Activity

Let's start stargazing! Just after sunset, go outside to observe the sky. You don't need any special equipment to do this, just your eyes. Can you find any constellations? Try to find Polaris. One way to find Polaris is to find the Little Dipper or Big Dipper first. The two stars of the Big Dipper's bowl point the way to Polaris. Look at the image below to see how this works. Record what you see in your science journal.

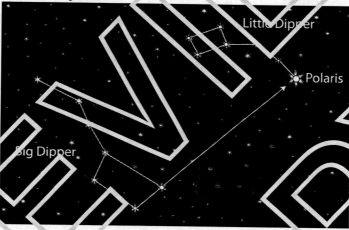

Journal Activity

Objective: Observe the difference between planets and constellations in the night sky.

Notes: Summarize what you know about planets and constellations.

Data: Find a constellation in the sky in the early evening. Try to find a planet. Then, look for them again a few hours later. Write the time of each observation and its location in your journal. Draw your own "map" of the night sky.

Summary: Describe how the constellation and the planet moved.

CHAPTER 1 REVIEW

1 **What is the main reason that stars appear to be so small when viewed in our night sky?**

    A   They are not evenly spaced.   C   They are very hot.

    B   They are very far away.   D   They are cooling off.

2 **A constellation is**

    A   a single star in a region of the sky.

    B   a grouping of stars in a region of the sky.

    C   the brightest star in the sky.

    D   a picture of a hunter.

3 **Which of the following is a galaxy?**

    A   Milky Way   C   Big Dipper

    B   Orion   D   Polaris

4 **The number of stars in the sky is in the**

    A   hundreds.   C   millions.

    B   thousands.   D   billions.

5 **Which of the following statements about stars is TRUE?**

    A   All stars in the sky are the same color.

    B   All stars in the sky are the same size.

    C   The shapes of constellations do not change.

    D   We feel heat from all stars in the sky.

Chapter 2
Observing Planets and Stars

INSTRUMENTS USED TO SEE THE STARS

You may be asking yourself how we know so much about stars if they are so far away. Well, most of what we know came from using just our eyes. The human eye was the first astronomical instrument! With just the eyes, it was possible to observe objects in the sky move, including the Earth.

People built instruments to help them see objects in the sky in more detail. One of the most basic tools we use is a pair of **binoculars**. Have you ever looked at something through a pair of binoculars? How did it look? Binoculars help us see very bright, large objects in the sky more closely. For example, it is possible to see the details of the Moon with binoculars. Binoculars are good tools to begin stargazing.

Figure 2.1 Binoculars

Scientists use more sophisticated tools, such as telescopes. A **telescope** is an instrument used to see dim and far objects much more closely. They work because objects in space give off light. Telescopes collect and focus this light. Scientists study it to understand the universe. We can see planets, stars, moons and other galaxies with telescopes.

Figure 2.2 Telescope

There are many types of telescopes. Some are positioned on the ground, and others are in space. The most famous is probably the Hubble Space Telescope. It is a telescope in orbit around the Earth. It collects clear images of distant objects in the sky.

Many more stars can be seen through a telescope than with just the eyes. Telescopes make stars and planets appear larger and brighter.

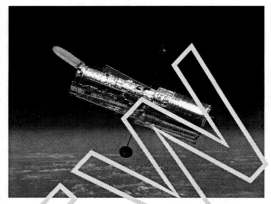

Figure 2.3 Hubble Space Telescope

Figure 2.4 Genesis Probe

Scientists use other instruments besides telescopes. They use probes to study objects in the Solar System. A **probe** is a spacecraft used for exploration. It is unmanned. This means it does not have a pilot or a crew. A probe collects and sends data to Earth for scientists. Probes can be sent far from Earth because they are unmanned. This makes them safer to use.

APPEARANCE OF STARS AND PLANETS

We have already learned a lot about stars. In the next chapter, we will learn a lot about the planets in our solar system. A **planet** is an object that is large enough to have gravity. Stars appear to the naked eye to twinkle on and off. Planets do not twinkle as much. Instead, they shine with a steady light. Planets usually appear brighter than stars because the planets are much closer to Earth. Planets closest to Earth are the brightest and easiest to see.

How can the difference between a planet and star be determined? One way is by how they look in the sky.

When viewed with a telescope or binoculars, planets look very different from stars. Planets are much smaller than stars. They are not as bright as stars. The shapes of planets and stars are different too. Planets look more like disks when viewed through a telescope. Stars are shaped like fuzzy balls of light. This is because stars are made of hot gases and produce their own light. Planets do not. They are made of substances that reflect light.

MOTION OF STARS VS. PLANETS

Stars are very far from Earth. Their movements are not noticeable to us on Earth. They appear to remain in the same positions relative to each other. So, you can think of the stars as a fixed background of lights in the night sky. The planets can be seen moving against the background of stars.

The planets we see are in our solar system. The solar system is our planetary neighborhood. The planets move around the Sun. Each planet moves at a different speed. As they move, they are seen in different places at different times when viewed from the Earth.

If you look at the night sky for several nights, you should be able to spot a planet by its movement. Because it is so close to Earth, it appears to move very fast. It changes its position in the sky each night. Look at Figure 2.5. This picture shows the motion of Mars over several months. You can see how it moves while the constellations (Pisces, Cetus, Taurus and Perseus) remain in place. A planet's movement can be seen using just the eyes.

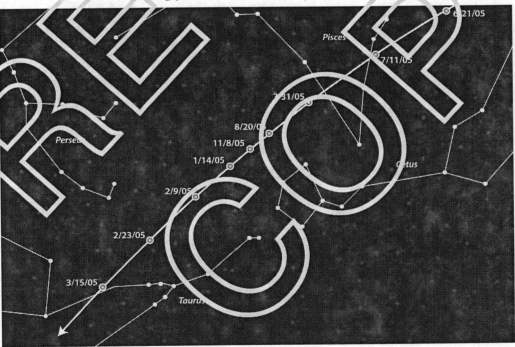

Figure 2.5 Motion of Mars

CHAPTER 2 REVIEW

1 What is a probe?

    A   a robotic spacecraft   C   a telescope

    B   a manned spacecraft   D   a dead star

2 Telescopes are instruments that

    A   send light into space to see objects.

    B   collect light from objects on Earth.

    C   collect different types of light from distant objects.

    D   collect only light that passes through Earth's atmosphere.

3 Why are probes sent really long distances?

    A   because they cannot see objects far away

    B   because they are unmanned

    C   because they are ground-based

    D   because they are in orbit around the Earth

4 Which of the following BEST describes the movement of the planets in the sky?

    A   They move around the Sun at different speeds.

    B   They move around the Sun at the same speed.

    C   They remain fixed.

    D   They stay in the same positions relative to each other.

5 Which of the following statements is TRUE?

    A   Stars are smaller than planets.

    B   Planets twinkle on and off.

    C   Planets produce their own light.

    D   Planets look like disks when viewed through a telescope.

Challenge Question

In the last fifty years, many new objects in our solar system have been discovered. In 2003 alone, twenty-five new objects (mostly moons of Jupiter, Neptune and Uranus) were discovered. Why do you think it took scientists so long to discover these objects? What do you think has increased scientists' ability to find objects in the solar system?

Chapter 3
The Solar System and the Moon

THE SOLAR SYSTEM

Our **solar system** is the part of our galaxy that moves around the Sun. The Sun is at the center of the solar system. Objects in our solar system **revolve** (move) around the Sun. The Sun's gravitational force keeps the planets revolving in a regular way. This **gravitational force** is the attraction that any object of mass has for other objects. The Sun's huge mass gives it a huge gravitational force. The force of gravity has a big impact on many aspects of the solar system.

Each object in the solar system revolves around the Sun in a slightly different way. Their paths are called **orbits**. The shape of an orbit depends on the size of the object. The orbit of a planet is slightly oval (almost circular). Other objects in the solar system have very oval orbits.

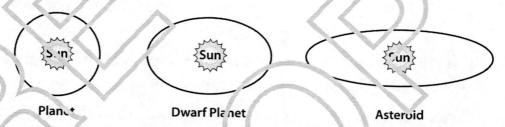

Figure 3.1 The Orbit of Objects around the Sun

PLANETS AND OTHER OBJECTS

Large objects in the solar system are called planets. Objects are called **planets** when they have enough mass to have their own gravity. The force of gravity makes a planet more round over many millions of years. **Dwarf planets** are much smaller than other planets in the solar system. Their gravitational force is also smaller. This means they are less round than familiar planets like Earth and Mars. In 2008, scientists developed a new category for objects in the solar system, **plutoids**. This is based on planetary shape and orbit. Plutoids must be round in shape. They must also revolve around the Sun in an orbit beyond Neptune. Pluto and Eris are now considered plutoids.

The planets in our solar system are shown in Figure 3.2.

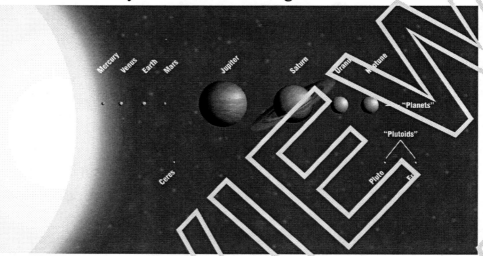

Figure 3.2 Solar System.

Looking at this figure, you should notice three important things:

1   **The Sun is much larger than anything else in our solar system.** It has the most mass. This mass means it has the largest amount of gravitational pull in the solar system. The result is everything in our solar system revolves around it.

2   **Our solar system has eight planets.** Pluto and Eris are plutoids. Ceres is a unique object and may be the only one of its kind.

3   **The first four planets are much smaller than the last four planets.** The inner planets, the first four, are called **terrestrial planets** (Mercury, Venus, Earth and Mars). The outer planets, the last four, are called **Jovian planets** or **Gas Giants** (Jupiter, Saturn, Uranus and Neptune).

Activity
Use Figure 3.2 to make a model of the solar system. Be sure your model correctly shows the size, location and the twelve recognized objects in our solar system. You can use objects like marbles, ping pong balls, tennis balls, pasta, foam or wads of paper for your model.

Journal Activity

Objective: To learn about the planets in our solar system.

Notes: Pick one planet in the solar system or the Sun. Use the library or Internet to gather information about your object. Get a recent picture of your object from the Hubble telescope or for the Sun, use soho (solar and heliospheric observatory). Use http://sohowww.nascom.nasa.gov, http://photojournal.jpl.nasa.gov/index.html or www.nasa.gov to get great pictures. Collect at least ten facts about your object. Include facts about size, orbit and number of objects that revolve around it.

Data: Use your facts to write a fan letter to your planets or the Sun. Ask for its autograph on your picture. Tell it why you like it the best. Don't forget to write its address at the top of the page like a real letter.

Summary: Share your letter with one of your classmates. Write a summary of your classmate's letter and what you learned about his or her planet or the Sun.

Activity

Science changes a lot. There are many times when scientists do not agree. This is part of the nature of science. Many scientists disagree over what to call Pluto. When Pluto was discovered in the 1930s, it was called a planet. Then it was a dwarf planet. Now, it is called a plutoid.

In small groups, discuss what you think Pluto should be called. Why do you think scientists need to change the classification of objects in the solar system? What factors might scientists use to classify objects? What role do scientific instruments (like telescopes and probes) play in the classification of objects in our solar system?

Challenge Question

Jupiter has 63 moons. Saturn has 60 moons. Mercury and Venus both have zero moons. Why do you think the Gas Giants have more moons than the terrestrial planets?

PHASES OF THE MOON

Most of the planets in our solar system have moons. **Moons** are smaller objects that revolve around a planet. Moons revolve around the planet they are closest to. They are smaller than their planet. They have been caught by the gravitational pull of that planet. Some planets have no moons. Some planets have many. For example, Jupiter has 63 moons! Earth's moon, **Luna**, is our only moon.

As the Moon revolves around the Earth, its shape appears to change. These changes are called **phases of the Moon**. The phases of the Moon are produced by the *alignment* of the Moon and the Sun in the sky. They are NOT caused by Earth's shadow. This is a common misunderstanding. Earth's shadow causes eclipses. An eclipse is seen when our view of the Moon or Sun is blocked. The Earth's shadow has nothing to do with the Moon's phases.

The side of the moon facing the Sun reflects sunlight. This is why the Moon is illuminated. The part of the Moon visible to us depends on the Moon's position relative to Earth.

On its journey from new moon to full moon, the moon is **waxing**. This is just a word that means it looks bigger. It is moving farther away from the Sun. We can see more of its surface because it is further from the Sun, and more light is being reflected. As it moves from full moon back to new moon, it is **waning**. This means it looks smaller. It is moving closer to the Sun. We see less of its surface because the angle between the Sun and Moon is very small and less light is reflected.

Journal Activity

Objective. Observe the phases of the Moon.

Notes. Name the four major phases of the Moon.

Data. Look at the Moon every night for 2 weeks. Write the date and draw a picture of the Moon next to the date each night. Write the name of the phase.

Summary: Describe the sequence of the phases with words. Record how long it took for the Moon to move between phases.

The position of the Sun and Moon determines the phase of the Moon. Follow Table 3.1 and Figure 3.3 as we describe the positions of the Moon and Sun for the four major phases.

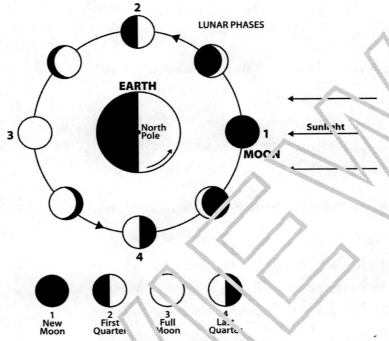

Figure 3.3 Phases of the Moon

Table 3.1 Phases of the Moon

Position	1	2	3	4
Phase	new	first quarter	full	third quarter
Visibility	tiny crescent	half moon	full	half moon

The angle between the Sun and Moon changes with each phase. It is smallest at new moon. This is position 1 in Figure 3.3. As you can see at new moon, the moon is almost directly between the Sun and the Earth so no light can be reflected for us to see. At first quarter (position 2), the Moon is half full. It has moved 1/4 of the way around its orbit.

At full moon (position 3), the Moon shines the brightest and is completely lit. It is the farthest from the Sun and is 1/2 of the way around its orbit. You can see from Figure 3.3, the entire surface of the Moon is exposed to sunlight. So, lots of light is reflected. Finally, at last quarter (position 4), the moon is 3/4 of the way around its orbit. Again it appears half full in the sky.

The Moon increases its brightness from right to left. It increases until the Moon reaches full. Then, the lighted part decreases from right to left until new moon. There is about a week between each major phase of the Lunar Cycle. So, there are 7 to 8 days between new moon and first quarter, between first quarter and full, and so on.

CHAPTER 3 REVIEW

1   The phases of the Moon are caused by the

A   Earth's shadow.

B   Sun's shadow.

C   alignment of the Sun and the Moon.

D   alignment of the Sun and Venus.

2   What causes the Moon to be illuminated?

A   light that is reflected from the Sun

B   light that comes from the Earth

C   light that comes from inside the Moon

D   light that is reflected from the stars

3   Why do the planets revolve around the Sun?

A   because our solar system is flat

B   because the planets are the same distance from the Sun

C   because the Sun's magnetic field attracts the planets

D   because the Sun is the largest object in the solar system

4   How are the four planets closest to the Sun physically different from the outer four planets?

A   The four planets closest to the Sun are made of gases and are very small.

B   The four planets closest to the Sun are made of rock and are very large.

C   The four planets closest to the Sun are made of rock and are very small.

D   The four planets closest to the Sun are made of gases and are very large.

5   If the lit part of the Moon is getting larger, it is

A   waxing.

B   waning.

C   at full phase.

D   at third quarter.

Chapter 4
Rotation and Revolution

How the Earth moves in space may seem unimportant. You might think it doesn't affect you very much. But really it affects you every day and every night. You see, how the Earth moves in space determines days, nights and our seasons. This is because of the Earth's relationship with the Sun.

ROTATION CAUSES DAY AND NIGHT

The Earth **rotates** (spins) around an imaginary line called an axis. The **axis** goes from the North Pole through the Earth to the South Pole. This is similar to a spinning top. If you have ever spin one around, you probably noticed that it spins around on a point, or on its axis. As it spins, it wobbles. This is a lot like the Earth's rotation on its axis.

Figure 4.1
Spinning Top

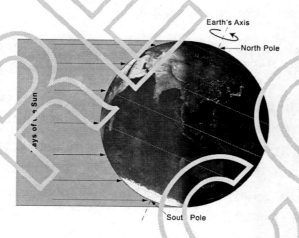

Figure 4.2 Day and Night

The shape of the Earth is a sphere. As the Earth rotates, different parts of the Earth face the Sun at different times. The part of the Earth that is facing the Sun is experiencing daytime. The part that is facing opposite the Sun is in darkness. This part is experiencing nighttime. The locations experiencing day and night change as the Earth rotates. It takes the Earth 24 hours to make one complete rotation. This is why one day is equal to 24 hours.

REVOLUTION CAUSES SEASONAL CHANGES

The Earth **revolves** in a fixed orbit around the Sun. It takes the Earth 365 days to make one trip around the Sun. This is why one year is equal to 365 days.

Earth's axis is not straight up and down. It is inclined, or tilted. The axis always points toward the North Star. The tilt of the axis causes the Sun's rays to strike the surface of the Earth at different angles. This is why we experience different seasons in a year. If Earth's axis were not inclined, there would be no change of seasons.

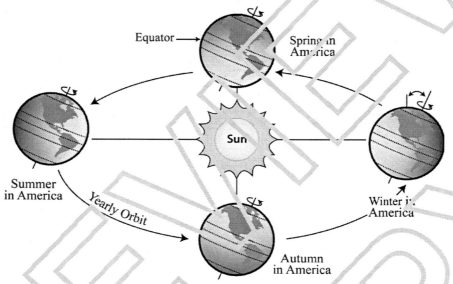

Figure 4.3 Sun/Earth Relationship

Figure 4.3 shows the relationship between the Earth and the Sun throughout the year. Notice how Earth's axis always points in the same direction in its orbit.

Sometimes the Earth's axis is tilted toward the Sun. This is when it is summer in the Northern Hemisphere. During summer, the hours of daylight are longer than the hours of nighttime. The increased sunlight makes temperatures warmer. It is hotter during the summer than the winter. At the same time, the Southern Hemisphere is pointing away from the Sun. The decreased sunlight means cooler temperatures. It is experiencing winter.

When the axis is tilted away from the Sun, it is winter in the Northern Hemisphere. During winter, the hours of nighttime are longer than the hours of daytime.

When the Earth is halfway between summer and winter in its revolution, it is fall or autumn in the Northern Hemisphere. When it is halfway between winter and summer, it is spring in the Northern Hemisphere. During fall and spring, the length of day and night is nearly equal.

Challenge Question

Look again at Figure 4.3. When the Northern Hemisphere is tilted away from the Sun, the Southern Hemisphere is pointing toward the Sun. What season is it in the Southern Hemisphere?

Journal Activity

Objective: Use sunrise and sunset to observe Earth's rotation.

Notes: Summarize what you know about Earth's rotation and day and night.

Data: Find a point in your yard and record the place of sunrise and sunset relative to this point. Do this every day until you start to see a pattern.

Summary: Describe how the location of sunrise and sunset changes over time.

Journal Activity

Objective: Observe the seasons on Earth where you live.

Notes: Summarize what you know about Earth's revolution and its axis to explain the seasons.

Data: In a table, record the average temperature and precipitation in each season for your hometown. Use the Internet to find your data, then plot the data in a graph.

Summary: Write a short paragraph describing each season in your hometown.

CHAPTER 4 REVIEW

1 **When the Earth's axis is tilted away from the Sun in Georgia, the season there is**

    A   summer.   B   winter.   C   spring.   D   fall.

2 **The Earth _____ on its axis one time every 24 hours, producing day and night.**

    A   revolves   C   rotates
    B   rejuvenates   D   changes orbit

3 **The Earth _____ around the Sun one time every 365 days.**

    A   revolves
    B   rotates
    C   rejuvenates
    D   changes orbit

4 **Which of the following is the MOST important in determining our seasons?**

    A   The rotation of the Earth.
    B   The revolution of the Earth.
    C   The length of day and night.
    D   The tilt of the Earth's axis.

5 **What would happen if the Earth took twice as long to revolve around the Sun?**

    A   The length of day and night would be shorter.
    B   The length of day and night would be longer.
    C   The length of a year would be longer.
    D   The length of a year would be shorter.

Chapter 5
States of Water

STATES OF WATER

There is a large amount of water on the Earth. Water covers about 71% of the Earth's surface. There is also water in Earth's soil and in the air. Water can take different forms on Earth. In other words, water exists naturally in three physical states. Water can be found as a **solid**, a **liquid** or a **gas**.

Does this sound familiar? It should. Last year in science, you learned about the three states of matter. What were they? They were solids, liquids and gases. Right, but how does water fit in?

Figure 5.1 Water on Earth

Water is a solid when it is very cold. Ice and snow are examples of solid water. You see solid water in winter. Water exists as a liquid in many places. You see liquid water in the oceans, rivers and lakes. It also comes out of the faucet in your home. When water is a gas, it is invisible. It exists as a gas as water vapor in the air. As a quick review, list as many examples as you can of water in its three states in the space below.

Solid	Liquid	Gas

What makes water change from a solid to a liquid? It's heat. Adding and subtracting thermal energy or heat changes the temperature of water. When thermal energy is added, water is heated. It changes from a solid to a liquid and finally to a gas. When thermal energy is lost, it cools the water. Water changes from a gas to a liquid and then a solid. Figure 5.2 shows how water changes when it is heated and cooled.

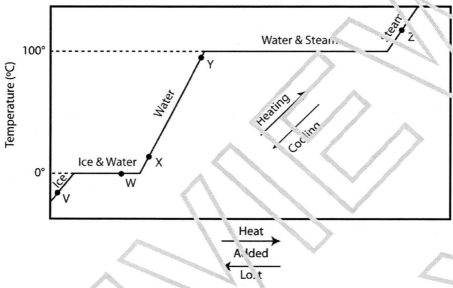

Figure 5.2 States of Water

When water changes states, it is called a **phase change**. There are four phase changes: freezing, melting, evaporation and condensation. During a phase change, the physical properties of water also change. It looks and behaves very differently in one phase compared to another.

Each phase change occurs at a specific temperature. The temperature can be measured using the Celsius or Fahrenheit scale.

Freezing changes liquid water into a solid. Solid water or ice is very hard. It has a definite shape and volume. Its particles are close together and are slow moving. Water freezes at 32°F or 0°C. Above 0°C, ice becomes a liquid. Freezing is the opposite of melting.

Melting changes solid ice into liquid water. Liquid water flows very easily. It has a definite volume but takes on the shape of its container. Ice melts at 0°C. Between 0°C and 100°C, water is a liquid.

Activity

Write down a few examples of melting and freezing in your science journal.

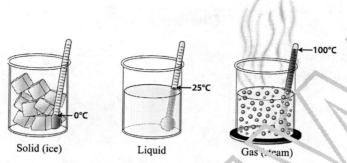

Figure 5.3 States of Water

Evaporation changes liquid water into a gas. You can see evaporation happening when water is heated to 100°C. At 100°C, liquid water becomes a gas. Water vapor has no definite volume. Its particles are far apart and fast moving. The bubbles that form in a pot of boiling water are pockets of steam. **Steam** is just another name for water in the gas phase.

Condensation changes steam into liquid water. It is the opposite of evaporation. Think of the pot of boiling water again. As the water in the pot boils, steam escapes into the air. Once in the air, it starts cooling to below 100°C. When it cools enough, it forms tiny liquid water droplets. They are so small they can float in the air. This makes a white smoky

Figure 5.4 Heat Transfer

cloud that we can see above the pot. The "white smoke" is a bunch of water droplets in the air formed from condensation. When many tiny floating droplets bump into each other, they eventually form heavier liquid water droplets. If you have ever boiled water in your kitchen, you might have noticed the condensation collecting on windows, cabinets or walls

---

Activity

Write down a few more examples of evaporation and condensation in your science journal.

Water is always changing between a solid, a liquid and a gas. These changes occur both on Earth and in the atmosphere. In Chapter 7, we will see how these three states are vital to all life on Earth.

Challenge Activity

Read each situation below. Decide what phase change is taking place. Write the phase change in the space to the right.

1. Droplets on the outside of a cold glass of water. _____

2. A sand castle in a sand pit falling apart after a day in the sun. _____

3. Ice cream dripping onto the floor that your dog licks up. _____

4. Blow drying your hair dry after a shower. _____

5. Raindrops changing to ice on the windshield of a car. _____

Journal Activity

Objective: Observe what happens to water as it changes between a solid and a liquid.

Notes: Summarize the phase changes of water and the temperatures at which they change.

Data: Fill a plastic cup with water. In a table, record the amount of water in your cup using a ruler. Place the cup of water in the freezer. When it is frozen, remove the cup. Record the time it took to freeze and the amount of ice in the cup. Use a ruler again to record the amount. Keep the cup out on the table. Let the ice melt. Record the time it took to melt and the amount of water now in the cup.

Summary: Describe what happened to the amount of water in the cup as it froze and melted. Did you notice anything about the outside of the cup as the ice melted?

Journal Activity

Objective: Draw a heating curve to show how water changes when it is heated.

Notes: List the three states of water and the temperature range for each state.

Data: Take a block of frozen water and heat it by placing it on a Bunsen burner or in a pan. Watch how the water changes. In a table, record the temperature of the water using a thermometer at 10 minute intervals. Write the state of the water in the table. Then, graph your data using a line graph.

Summary: Describe your graph in your own words. How did the water change as it heated?

Journal Activity

Objective: Identify objects that can freeze and melt.

Notes: Summarize the difference between freezing and melting.

Data: Make two columns in your journal, one for freezing and one for melting. Find objects that can freeze inside your house and outside in your yard. Write them in the freezing column. Then do the same for objects that can melt.

Summary: Describe the types of things you found. Were your columns the same or different?

Challenge Question

Draw three circles and label them solid, liquid and gas. In the circles, draw how the molecules look in each state of water. How packed together or spread apart are the molecules? Think about the properties of water in each state.

CHAPTER 5 REVIEW

1   Which of the following is an example of condensation?

   A   Ice cream dripping on the ground.

   B   Water droplets on the outside of a glass of ice tea.

   C   A snowman disappearing after a day in the sun.

   D   Frost or icicles forming when it is really cold outside.

2   At what temperature would liquid water become solid ice?

   A   100°C

   B   212°F

   C   32°C

   D   0°C

3   In which of the following situations would water have the highest temperature?

   A   when water is frozen as ice

   B   when water is boiling

   C   when water is melting to a liquid

   D   when water is condensing

4   Which phase change occurs when evaporation takes place?

   A   Water changes from a liquid to a solid.

   B   Water changes from a gas to a solid.

   C   Water changes from a liquid to a gas.

   D   Water changes from a solid to a gas.

5   Which temperature would you find liquid water?

   A   110°C

   B   –4°C

   C   0°C

   D   25°C

Chapter 6
Clouds

Have you ever thought a cloud looked like an animal or other object? Perhaps you once thought a cloud looked like a dragon. Or maybe a dinosaur. It's fun to imagine all the things a cloud can look like! Clouds come in all different shapes and sizes. And they bring different types of weather.

What *exactly* is a cloud?

That cloud looks like a cross between a rabbit and a bat It's a rabbat!

Figure 6.1 Rabbit or Bat?

Clouds are just liquid or solid water in the atmosphere. A **cloud** is a group of tiny water droplets that float in the atmosphere. Clouds are sometimes made of ice crystals. Winds blow and move the clouds. They move in the direction the wind is blowing. Before we look at different types of clouds, let's investigate how they get there in the first place.

Cloud Formation

Air must rise for clouds to form. Air rises when it is warmed. Heat from the Sun warms the air near the surface of the Earth. It also makes more water vapor through evaporation. As air is heated, it expands. This makes the air lighter. Cooler, denser air falls and forces the hot air to rise. Temperatures high in the atmosphere are much cooler than at the Earth's surface. This makes the water vapor condense. In addition, smaller particles like dust, pollen or pollution must be present to form a cloud. They act like a seed. They provide a surface for water vapor to attach to. During cloud formation, water changes from a gas back to liquid water droplets.

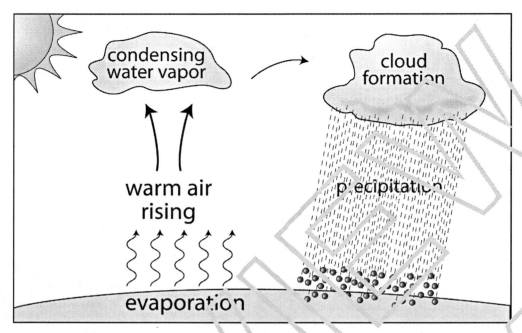

Figure 6.2 Cloud Formation

There are three basic ways that air rises to form clouds: convection, a frontal boundary and by elevation.

Convection is the movement of warm air upwards. It is triggered by heat from the Sun. The Sun strikes the surface of the Earth and warms the air near the ground. The warm air is less dense or lighter than cool air. Heavy, cool air nearby rushes in and forces the warm air up.

The second way is along frontal boundaries. A **frontal boundary** occurs when a warm air mass and a cold air mass meet. Warm air is lighter than cold air. So, the warm air is forced upwards. It rises and cools quickly, sometimes forming severe weather. We will learn more about air masses and fronts in Chapter 8.

The third way is by **elevation**, like a mountain. When warm air moves sideways over land or water, it sometimes meets a mountain. The mountain blocks the flow of air. The air cannot continue moving forward. It is forced to move up along the side of the mountain. As it climbs the mountain, it cools.

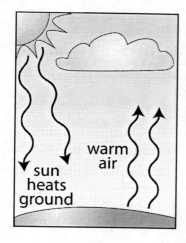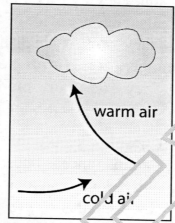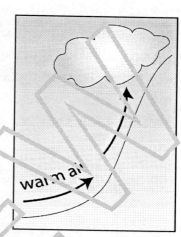

Figure 6.3 Lifting Mechanisms

TYPES OF CLOUDS

The names of clouds are taken from Latin. These names describe the cloud's height from the ground and its shape. By learning the meaning of just a few names, you can identify almost every cloud in the sky! So, let's get started.

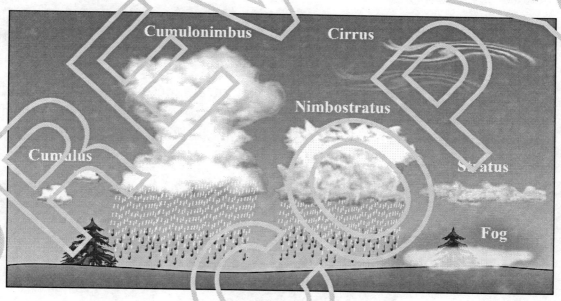

Figure 6.4 Cloud Types

Cirrus clouds are the highest of all clouds. Cirrus means a "curl of hair" in Latin. They are very thin and are described as wispy clouds. They are made of tiny ice crystals. This is because they form very high in the sky where temperatures are very cold. Cirrus clouds mean calm weather. They do not produce precipitation. They are too thin.

Figure 6.5 Cirrus Clouds

Cumulus clouds are mid and low level clouds. Cumulus means "heap" or "pile" in Latin. They are puffy clouds and look like puffy balls of cotton in the sky. Cumulus clouds are fair weather clouds. Small cumulus clouds that are scattered across the sky mean good weather. If they grow bigger, they may produce precipitation.

Figure 6.6 Cumulus Clouds

The lowest clouds in the sky are **stratus** clouds. These are flat sheets of clouds that form in layers. Stratus means "to spread out" in Latin. They usually cover the entire sky and bring overcast, gray weather. Stratus clouds do not always produce precipitation. When they do, it is usually light rain. Stratus clouds can produce snow when the air is cold enough.

Figure 6.7 Stratus Clouds

A **nimbus** cloud always produces precipitation. Nimbus means "rain" in Latin. Its name is combined with other cloud names. For example, nimbostratus is a stratus cloud that forms precipitation. A cumulonimbus cloud is a cumulus cloud that forms precipitation.

A cloud that touches the ground becomes **fog**. It is usually a stratus cloud that touches the ground. Fog forms when air near the surface is very moist and cools enough for condensation to occur. Fog can be thin (less dense) or very thick (dense). Dense fog makes it hard to see. This can be very dangerous, especially for drivers!

Figure 6.8 Fog

TYPES OF PRECIPITATION

Figure 6.9 Cumulonimbus Clouds

When water droplets in the clouds bump into each other, they get bigger. As they get bigger, they get heavier. Eventually, they become too heavy to stay in the air. They fall to the ground as rain or snow. Rain and snow are two types of precipitation. **Precipitation** is water falling from the sky or atmosphere. **Rain** is liquid water that falls as separate drops. Rain does not look like a teardrop. Instead, it looks more like a sphere or small beach ball. **Snow** is solid ice crystals that fall to the ground.

Hail is another type of precipitation. It is a ball or lump of ice. Hail forms when water droplets in the cloud move up and down inside the cloud, melting and refreezing. This builds layers of ice. Hail can be small like the size of a pea or as big as a baseball. Cumulonimbus clouds can make hail during a thunderstorm.

Figure 6.10 Snowflake

When the temperature is near freezing, precipitation may fall as sleet. **Sleet** is frozen raindrops. It is like an ice pellet. Sleet forms as raindrops fall from a cloud and pass through freezing air. The raindrops freeze before hitting the ground.

Sometimes water droplets form on the surface of objects near the ground. This is called dew. Dew is not really a form of precipitation. It does not fall from the atmosphere. It forms as water vapor condenses into drops of water. The drops of water form on the surface of an object that is cool. These surfaces are things like blades of grass, leaves and cars. You may have seen dew on your lawn early in the morning.

Journal Activity

Objective: Observe the different types of clouds in the sky.

Notes: Summarize the main types of clouds.

Data: Draw a picture of each cloud that you see in the sky. Write the name of the cloud next to the picture.

Summary: Describe the clouds and the weather. Is there any precipitation? In what form?

Activity

Become a storm spotter! Forecasters can use clouds to predict the weather. Imagine you are a storm spotter helping a weather forecaster. It is your job to predict the weather and report it to the local forecaster. Match each cloud to the correct weather condition below.

Nimbostratus   Cumulus   Cumulonimbus   Cirrus

  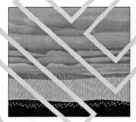 

Challenge Activity

Write the meaning of the Latin names for each type of cloud.

Latin	Meaning
Cumulus	
Cirrus	
Nimbus	
Stratus	
Cumulonimbus	
Nimbostratus	

Activity

What kind of severe weather do you get in your hometown? List the most common types that occur where you live.

List the steps to take to prepare for each kind of weather event. For help, use these Web sites: http://fema.gov/ and http://www.crh.noaa.gov/oax/safety/safety.php.

Use a piece of construction paper. Fold it in half or in thirds to create a brochure.

Add a title and a picture to the front page. On the inside, write the names of each weather event and the steps to prepare for each one.

You now have a severe weather safety guide. Share it with your friends and family. Make copies and distribute them around your community!

Journal Activity

Objective: Predict different forms of precipitation.

Notes: Summarize the forms of precipitation that you are familiar with.

Data: Identify the types of clouds in the sky. Predict if precipitation will develop. What form it will be in? Check your prediction. Remember to consider the temperature too. Do this for a few days to practice making predictions.

Summary: Write a paragraph summarizing your predictions. Were they right? If not, why do you think they were wrong?

Activity

Write a short story. The names of the characters will be the name of clouds. Describe what each character is like. Use what you know about the cloud. Maybe "stratus" is a gloomy, sad person. Maybe "cirrus" is cheery. Create an adventure for your characters!

CHAPTER 6 REVIEW

1   **Which of the following processes is NOT part of the development of a cloud?**

    A   precipitation   C   condensation

    B   evaporation   D   rising air

2   **Which of the following is NOT a form of precipitation?**

    A   rain   C   sleet

    B   hail   D   dew

3   **Hail is a type of frozen precipitation. It can fall from what type of cloud?**

    A   cirrus   C   nimbus

    B   cumulonimbus   D   stratus

4   **When a cold air mass meets a warm air mass, what happens?**

    A   Warm air forces the cold air upwards.

    B   Warm and cold air rise.

    C   Cold air forces the warm air upwards.

    D   Warm and cold air sink.

5   **What does the name of a cloud tell us?**

    A   The height of the cloud and what it looks like.

    B   The height that the top of the cloud reaches.

    C   The type of precipitation that it is producing.

    D   How it was formed.

Chapter 7
The Water Cycle

THE WATER CYCLE

All living things need water. Without it, living things will die. Most of the water on Earth is in the oceans. The oceans are made of salt water. There is much more salt water than fresh water. There must be a natural way to exchange the salt water for fresh water. There is, and we call it the Water Cycle! The **Water Cycle** moves fresh water between the atmosphere and Earth.

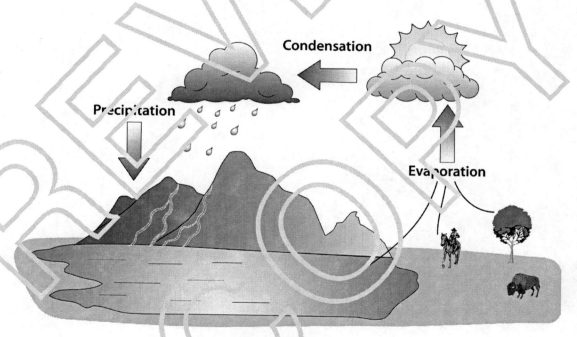

Figure 7.1 The Water Cycle

Heat from the Sun evaporates liquid water on Earth. This causes liquid water to become water vapor in the atmosphere. Water can evaporate from oceans, rivers, lakes, puddles or the ground. Even plants and animals contribute to water vapor in the atmosphere. When the water evaporates from ocean water, it leaves all the salt behind.

As warm air rises, it cools. Cooler temperatures cause the water vapor to condense into clouds. Eventually, the tiny droplets of water inside the clouds collect enough water to make rain. Remember in Chapter 6, we learned that the water droplets inside a cloud bump into each other. Precipitation then falls to the ground.

Water on the ground flows into rivers, lakes and eventually, the ocean. We call this **runoff**. This starts the cycle all over again.

Without this cycle of evaporation, condensation and precipitation, a fresh water supply would not be available. Without fresh water, you'd run out of energy very quickly.

Activity

Use the space below to draw your own water cycle.

Journal Activity

Objective: Observe how condensation forms.

Notes: Write a definition of condensation and its role in the water cycle.

Data: Fill a glass with water. Then, add several ice cubes to the water. Place the glass of ice water in a warm room or outside on a sunny, warm day. Watch what happens on the outside of the glass. Describe the changes you see in your journal.

Summary: What changes occurred? Why did they occur? What would happen if the glass of water was kept outside longer?

Journal Activity

Objective: Make your own water cycle.

Notes: Summarize main parts of the water cycle.

Data: Use a small plastic container, like a small shallow dish. Find soil from your yard. Put it in the container. Place a seed in the soil. You could use a corn or bean seed. Water the soil. Cover the container with clear plastic wrap. Place it next to a window in the Sun or outside. Record changes you see each day for 2–3 weeks.

Summary: Describe what happened to the plant. Talk about how the water moved around in your container. How is this like the water cycle?

Challenge Question

Evaporation takes place when water is heated. It occurs in fresh water and salt water. Salt is a product that people get by evaporation. How does evaporation change the amount of salt in water?

CHAPTER 7 REVIEW

1 **The water cycle circulates _____ between the Earth and the atmosphere.**

    A   fresh water   C   heat

    B   salt water   D   energy

2 **What heats the liquid water on Earth to create water vapor?**

    A   the Sun   C   the stars

    B   the Moon   D   plants and animals

3 **How is water moved from the surface of the Earth to the atmosphere?**

    A   condensation   C   freezing

    B   precipitation   D   evaporation

4 **Water vapor in the atmosphere does NOT come directly from**

    A   soil.   C   plants.

    B   rivers.   D   ground water.

5 **Which season would have the greatest amount of evaporation?**

    A   spring   C   fall

    B   summer   D   winter

Chapter 8
Weather Patterns

WEATHER VERSUS CLIMATE

Weather and climate are two terms often used together. However, there is a very important difference between them. This difference has to do with time. **Weather** refers to daily or hourly changes. It describes the atmosphere at a specific time and place. Sunny, windy, cloudy, stormy… these are terms we use to describe the weather.

Climate is the usual weather over a long period of time in a region. Think of it as the weather average. There are many different climates in the world. Some examples are desert, tropical and polar climates. Georgia has a subtropical climate. It has hot, humid summers and cool winters. Georgia gets a lot of precipitation all year. Some years are wetter than others

Figure 8.1 Desert   **Figure 8.2** Tropical Rainforest

WEATHER FRONTS

Changes in the weather happen as different air masses move in and out of a region. **Air masses** are large bodies of air. They have the same temperature, moisture and pressure. Think of air masses as "blocks" of air that move in the atmosphere. They cause most of the weather that we experience.

The movement of one air mass affects another air mass. This forms a **front**. Fronts are boundaries that separate different air masses. There are two basic kinds:

- **Cold Front** – The leading edge of colder air that is replacing warmer air.
- **Warm Front** – The leading edge of warmer air that is replacing cooler air.

The movement of air masses creates different weather. When one air mass moves out of a region, another air mass moves in. If the two air masses are really different, severe weather can form.

WEATHER MAPS

Weather is an important part of our daily lives. It often affects our activities. A prediction of the weather for a future time is called a **forecast**. Forecasts help us plan what to wear.

To make a forecast, scientists must analyze a lot of data. To compare lots of data more easily, they use **weather maps**. A weather map helps forecasters predict the weather. It also displays important information to the public. You might have seen one on the evening news. Many different types of weather maps exist. You are probably familiar with some already, like radar maps and temperature maps.

Maps that show a lot of information use **weather symbols**. A weather symbol can be a picture or a letter. They show complex information in a clear and easy way. Let's look at some common weather maps.

COLD FRONT

A **cold front** is a cold air mass that replaces a warmer air mass. The air behind a cold front is much colder and drier than the air ahead of it. On a weather map, a cold front is a solid line with triangles. The triangles point towards the warmer air. They also point in the direction of movement. Figure 8.3 shows the symbol for a cold front.

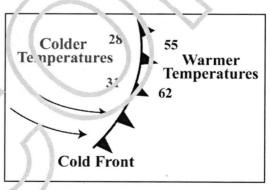

Figure 8.3 Cold Front

WARM FRONT

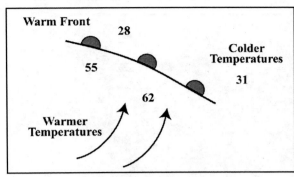

Figure 8.4 Warm Front

A **warm front** is a warm air mass that replaces a cold air mass. The air behind a warm front is warmer and more moist than the air ahead of it. On a weather map, a warm front is a solid line with semicircles. The semicircles point towards the colder air and in the direction of movement. Figure 8.4 shows the symbol for a warm front.

PRESSURE

Forecasters use air pressure to monitor changes in the weather. **Air pressure** is the weight of air on the surface of the Earth. When there is more air above an area, the pressure is higher. When there is less air above an area, the pressure is lower. There are two types of pressure systems, high pressure and low pressure.

High pressure occurs when the air pressure is greater than surrounding areas. The symbol for high pressure is an 'H." The "H" is blue on a weather map. It shows where the center of a high pressure air mass is. This is where skies are clear.

Figure 8.5
High Pressure

Low pressure occurs when the air pressure is lower than surrounding areas. The symbol for low pressure is an "L." The "L" is red on a weather map. It shows where the center of a low pressure air mass is. This is where it is cloudy and rainy.

Figure 8.6
Low Pressure

The weather map in Figure 8.7 shows fronts, air pressures and precipitation for the United States.

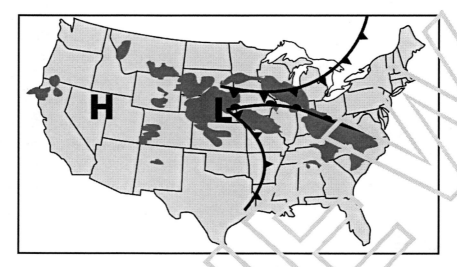

Figure 8.7 U. S. Weather Map

There are three fronts on the map. Can you identify all three?

There are two cold fronts. One is moving from the north to the south. The other is moving from the west to the east. There is also a warm front. It stretches from east to west across the east central United States.

The shaded parts of the map show where there is precipitation. This may be in any form, such as rain, snow, or hail. The "L" shows the center of the low pressure. This marks the center of a storm. Here, there is precipitation and it is very wet. Notice that it is mostly dry around the center of high pressure.

Challenge Question

Write a short forecast for Georgia using Figure 8.7. Describe the weather now and how it will change in the near future.

Journal Activity

Objective: Observe and interpret weather maps.

Notes: Summarize what you know about weather maps.

Data: Draw the daily weather map in your journal. Use the Internet or a local newspaper to view the weather map. Next to the map, record the high and low temperature, precipitation and air pressure for your city. Do this for one week.

Summary: Use your maps and data to describe any patterns you see. Make a forecast for the next day. Check your prediction.

Activity

Label the front, temperature and precipitation on each map. Use the space next to the map to describe the weather in Atlanta, GA.

_____

_____

_____

_____

_____

_____

CHAPTER 8 REVIEW

Use the figure to answer questions 1 and 2.

1   **Which of the following MOST LIKELY describes the weather in Atlanta?**

    A   warm and dry   C   cold and dry

    B   warm and rainy   D   cool and rainy

2   **Which of the following MOST LIKELY describes the weather in Savannah?**

    A   warm and dry   C   cold and dry

    B   warm and rainy   D   cool and rainy

3   **What does an "L" on a weather map mean?**

    A   high pressure and clear skies   C   low pressure and clouds

    B   low pressure and clear skies   D   cold and rainy weather

4   **A front forms when**

    A   one air mass moves another air mass.

    B   one air mass moves.

    C   two air masses mix and form a new air mass.

    D   two air masses do not move.

5   **When the difference between two air masses is large, the weather is**

    A   less severe.   C   less predictable.

    B   more severe.   D   more predictable.

Chapter 9
Instruments in Weather Forecasting

WEATHER INSTRUMENTS

Scientists need a lot of data to make a forecast. They collect data using weather instruments. **Weather instruments** measure and record conditions of the atmosphere. These measurements tell us what the weather is like. Scientists monitor the data to find patterns. Patterns in the weather help them to make predictions. Different instruments are used to gather weather data

TEMPERATURE

Do you know what instrument measures temperature? If you said a thermometer, you are right! A **thermometer** is a closed glass tube. It is filled with alcohol or mercury. Air around the tube heats the liquid inside. This causes the liquid to expand and move up the tube. A scale on the tube shows the temperature. The two most common scales are Celsius and Fahrenheit.

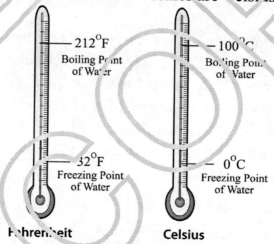

Figure 9.1 Thermometers

Temperatures affect every living thing. Changes in temperature cause different weather. Sometimes even the slightest change in temperature creates very different weather. If it drops below freezing, rain turns into snow.

PRESSURE

Air pressure can also change with the weather. It rises and falls as air masses move. This makes it very useful in forecasting. Air pressure is measured with a barometer.

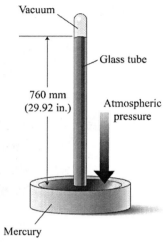

Vacuum

Glass tube

760 mm
(29.92 in.)

Atmospheric
pressure

Mercury

Figure 9.2 Barometer

A **barometer** tells us if the pressure is rising or falling. A rising barometer means drier weather. A falling barometer means cloudy and wetter weather. A barometer is filled with water or mercury. As the pressure changes, the water or mercury level changes. This change is recorded. Air (or barometric) pressure is typically measured in inches of mercury (in Hg).

Barometers tell us where there is high and low pressure. Areas with high pressure have calm weather. Areas of low pressure have changing weather. The change in pressure is most important. Rising pressure means a high pressure system is moving into the area. What type of weather is associated with high pressure? Right! It's calm, dry weather. Falling pressure means a low pressure system is moving in. What type of weather is associated with low pressure? You're right, Rain! This is how forecasts are made.

PRECIPITATION

A **rain gauge** measures the amount of rain that falls in one place. This instrument is an open cylinder. It sometimes has a funnel at the top to help catch rain. It has a small measuring tube inside of it. This measures the total rainfall. Rainfall is measured in millimeters (mm) or inches (in). They are placed outside in open areas to catch the rain that falls to the ground.

A rain gauge measures rainfall for a specific place. Several rain gauges averaged together tell us the total rainfall in an area. They are not always reliable though. Objects can fall into the rain gauge or on top of it. If this happens, measurements are less reliable.

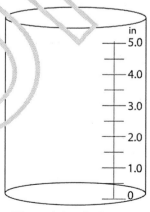

in
5.0
4.0
3.0
2.0
1.0
0

Figure 9.3 Rain Gauge

WIND

A **weather vane** (or wind vane) is an instrument that measures wind direction. Weather vanes are placed on high objects that are more exposed to the wind. A rooftop is a good place for a weather vane. Sometimes the top of a weather vane is shaped like an animal. The wind blows and turns the weather vane. Arrows on the weather vane show the direction of the wind. A weather vane points in the direction the wind comes from. For example, let's say the wind blows from east to west. The weather vane points toward the east.

Figure 9.4 Weather Vane

Wind speed is measured with an **anemometer**. This might be a hard word to pronounce, but it is a simple instrument. An anemometer has cups attached to the end of four horizontal arms. The cups catch the wind and turn. A dial that is attached to it tells the wind speed. Winds are light when the weather is calm. Winds are strong during stormy, severe weather.

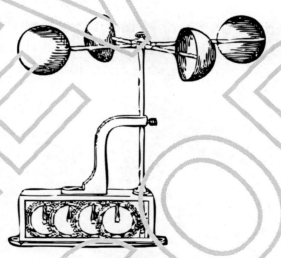

Figure 9.5 Anemometer

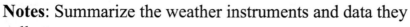

Journal Activity

Objective: Record and graph weather data.

Notes: Summarize the weather instruments and data they collect.

Data: Choose four different weeks out of the year, one in each season. In a table, record the high and low temperatures, wind speed and direction, and the precipitation in your city. Do this for each day of the week. Use the Internet, a local newspaper or a TV weather report. Graph your data. Make one graph for each week.

Summary: Describe the weather for each week. Compare the four graphs and describe the different seasons.

Activity

Look at the barometers below. Order them from highest to lowest. Write numbers 1 (lowest) to 3 (highest) in the spaces provided. Draw a line connecting the barometer to the correct weather condition.

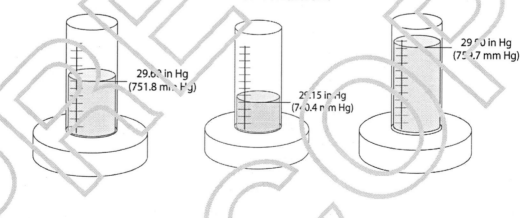

29.60 in Hg
(751.8 mm Hg)

29.15 in Hg
(740.4 mm Hg)

29.90 in Hg
(759.7 mm Hg)

_____   _____   _____

sunny and dry   partly cloudy with showers   cloudy with thunderstorms

Activity

It has been raining all day! You want to know how much it has rained. You run outside to look at your rain gauges. Color the empty rain gauges to show how much rain fell. Use a green colored pencil or crayon. The level of rain is given below each rain gauge.

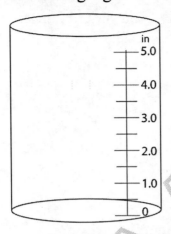
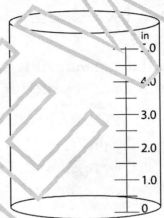

Rainfall total at your house 1.3 inches   Rainfall total at your school: 3.4 inches

Why were there different rainfall amounts in different locations (house vs school)?

When you got home, you noticed a leaf had fallen on top of your rain gauge. What does this mean? Which rain gauge measurement is more reliable? Why?

Challenge Activity

The table shows weather data for yesterday and today in Atlanta, Ga. Write the forecast for tomorrow in your own words. You know tomorrow the pressure will continue falling. The winds will continue to rise and be from the north. Use today's weather to talk about how the weather will change tomorrow.

Weather Data	Yesterday	Today
Temperature	72°F	60°F
Precipitation	0.02 in	1.20 in.
Pressure	29.90 in Hg	29.55 in Hg
Wind Speed	2–5 mph	10–12 mph
Wind Direction	SW	NW

CHAPTER 9 REVIEW

1 **A(n) _____ is used to measure wind speed, and a(n) _____ is used to measure wind direction.**

    A   weather vane, thermometer   C   anemometer, weather vane

    B   psychrometer, barometer   D   barometer, anemometer

2 **In the morning, it is partly cloudy. By the afternoon, it is completely calm and very sunny. You conclude that**

    A   the pressure is rising.

    B   the pressure is falling.

    C   the pressure has remained the same.

    D   the pressure fell but is now rising.

3 **A falling barometer indicates what type of weather?**

    A   rain

    B   calm

    C   sunny

    D   high pressure system

4 **The wind is moving from south to north. Which way is the weather vane pointing?**

    A   north

    B   west

    C   south

    D   east

5 **In the springtime, you move your house plants outside. Your plants cannot survive freezing temperatures. What weather instrument do you need to help you know when to bring your plants back indoors?**

    A   anemometer   C   wind vane

    B   thermometer   D   barometer

Domain One Review
Earth Science

1 **What causes the movement of constellations in the night sky?**

    A   their movement around the Earth

    B   the Earth's movement around the Sun

    C   the movement of the stars around the Sun

    D   their movement around the stars

2 **How are stars alike?**

    A   They are all white.

    B   They all have the same brightness.

    C   They all produce their own light.

    D   They are the same size.

3 **Why do scientists use instruments to view stars and planets?**

    A   because they can't be seen with the eyes

    B   to make them appear smaller

    C   to count how many there are

    D   to see them more clearly

4 **Nadia lives in Atlanta, Ga. It is spring. How many hours of day and night is she experiencing?**

    A   about the same hours of day and night

    B   more hours of day than night

    C   less hours of day than night

    D   You cannot tell from the information given.

5 It is a really cold day. Cindy goes outside and breathes out. She sees her breath as a white cloud. What causes this?

A   evaporation

B   freezing

C   condensation

D   deposition

6 A cirrus cloud moves across the sky. It becomes thinner and then disappears. What caused it to disappear?

A   it lost all its energy

B   it evaporated

C   it condensed

D   it moved higher up in the sky and was no longer visible

7 What type of weather system is represented with a solid line and triangles?

A   low pressure

B   high pressure

C   warm air mass

D   cold air mass

8 Kenda listens to the weather forecast. It says there will be high pressure tomorrow. What will the weather probably be like?

A   cloudy and rainy

B   sunny and dry

C   very stormy

D   showery and windy

Use the image to answer question 9.

9 What types of clouds are shown in the picture?

A   cumulus and stratus

B   cumulus and nimbus

C   cirrus and stratus

D   cirrus and cumulus

10 The Earth's axis is tilted away from the Sun. What season is it in Atlanta, GA?

A   winter   B   spring   C   summer   D   fall

Domain Two
Physical Science

Chapter 10: How Light Behaves

S4P1a: Identify materials that are transparent, opaque and translucent.

S4P1b: Investigate the reflection of light using a mirror and a light source.

Chapter 11: White Light

S4P1c: Identify the physical attributes of a convex lens, a concave lens and a prism and where each is used.

Chapter 12: Sound

S4P2a: Investigate how sound is produced.

S4P2b: Recognize the conditions that cause pitch to vary.

Chapter 13: Work and Simple Machines

S4P3a: Identify simple machines and explain their uses (lever, pulley, wedge, incline plane, screw, wheel and axel).

Chapter14: Force and Motion

S4P3b:Using different size objects, observe how force affects speed and motion.

S4P3c: Explain what happens to the speed or direction of an object when a greater force than the initial one is applied.

S4P3d: Demonstrate the effect of gravitational force on the motion of an object.

Chapter 10
How Light Behaves

HOW LIGHT BEHAVES

Light travels fast. It travels so fast that we do not even think of it as moving. Instead, we think of it as just being there.

Let's say we shine a flashlight on some surface, like a wall. How does the light travel from the flashlight bulb to the wall's surface? Let's investigate it step by step. When you turn the flashlight on, light comes out of the bulb (point A). It moves out into the air (point B). Finally, it reaches the surface (point C).

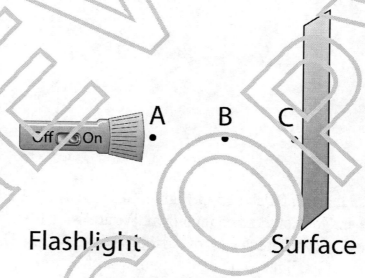

Figure 10.1 Light Travels to a Wall

What happens when the light hits the surface?

Light that hits a surface can do one of four things:

1   It can go into the surface. We say that the light has been **absorbed** by the surface. If the surface absorbs the light, it stops moving at point C.

2   It can go through the surface. We say that the light has been **transmitted** by the surface (point D).

3   It can go through the surface bent. We say that the light has been **refracted** by the surface (point E).

4   It can bounce off the surface. We say that the light has been **reflected** by the surface (point F).

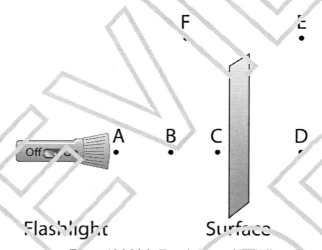

Figure 10.2 Light Travels Through a Wall

You see the points labeled in Figure 10.1 and 10.2? Let's take them out of the picture and look at them on their own. We are using these points to make a model. A **model** is a different way of looking at something. This model describes the behavior of the light as it comes out of a flashlight.

LIGHT GETS ABSORBED

To build our model, think of the light moving from point A to point B to point C. The light makes a straight line. At point C, we will show the surface as a straight line. If the surface at point C absorbs the light, the line stops there.

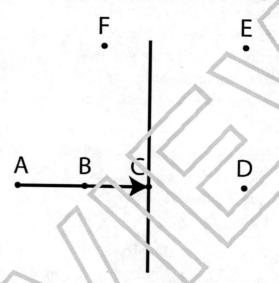

Figure 10.3 Light is Absorbed

LIGHT GETS TRANSMITTED

If the surface transmits the light, the line can continue past point C. If it goes straight through, without being *refracted*, it will come to point D. Refracted means bent. Light that is transmitted has a zero angle of refraction. The **angle of refraction** measures how much light gets bent.

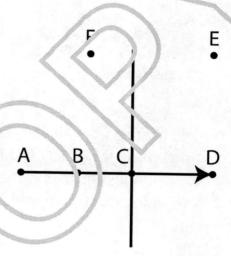

Figure 10.4 Light is Transmitted

LIGHT GETS REFRACTED

Usually when light moves from one material through another, it gets refracted or bent in some way. This changes the way it was traveling. The amount that the light is bent is called the angle of refraction. In Figure 10.5, the angle of refraction is shaded for you to see more easily. In this example, after the light passes through the material, it is bent. It then travels to point E.

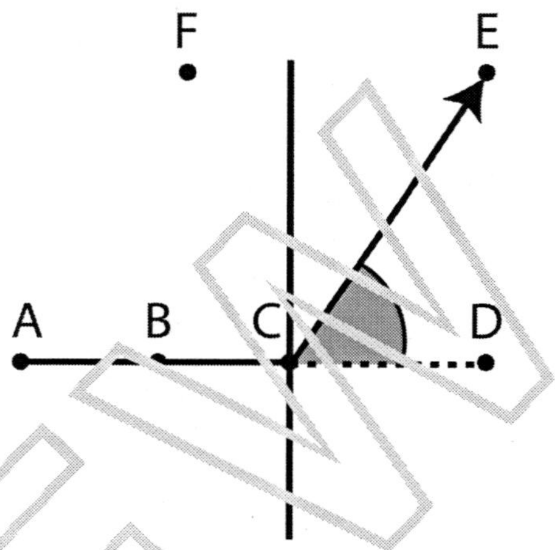

Figure 10.5 Light is Refracted

The angle of refraction depends on several factors including the speed of the light. As the light passes through a material, its speed is determined by the density of the material it is traveling through.

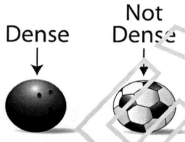

Figure 10.6 Comparing Density

Density is a way to measure how much matter is packed into an object. Dense objects have large amounts of matter packed tightly together. A bowling ball is very dense. It is also heavy. Less dense objects are made of matter that is more spread out. A soccer ball is not dense. It is very light! Both a bowling ball and a soccer ball are about the same size. But their densities are very different.

Light passing from air (not dense) through a glass of water is slowed by the water. The light is bent or refracted as it moves through the denser (dense) water. This is why the straw in Figure 10.7 appears broken.

Figure 10.7 Refracted Straw

LIGHT GETS REFLECTED

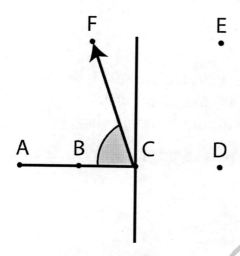

Figure 10.8 Light is Reflected

When light hits a surface, it may not allow the light in at all. In that case, it bounces back the way it came. We say that the light is reflected. A shiny surface like aluminum foil reflects light. The angle that the light is bent is called the **angle of reflection**. In Figure 10.8, the light ray strikes the surface. Then it bounces off. It then travels to point F. The angle of reflection is shaded for you to easily see.

A NOTE ON ANGLES

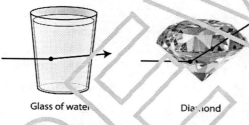

Glass of water   Diamond

Figure 10.9 Density Affects Refraction

The more the light is bent, the larger the angle of refraction is. Denser objects bend light the most. Water is not dense. It is a liquid. The particles that make up water are spread out. On the other hand, a diamond is very dense. It is a solid. Its particles are tightly packed. Diamonds bend the light more than water. They have a large angle of refraction. Water has a small angle of refraction.

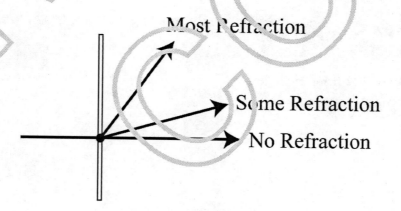

Most Refraction

Some Refraction

No Refraction

Figure 10.10 Refraction Angles

Sometimes a single beam of light can be absorbed, refracted and reflected, all at the same time. We'll talk more about this in the next chapter.

THE SURFACE MATTERS

How do you know what light will do once it hits the surface? It depends on what the surface is made of. Different types of material affect light differently. That is because light travels at different speeds in the different materials.

Opaque materials do not allow light to pass through them. Opaque materials either absorb light or reflect it.

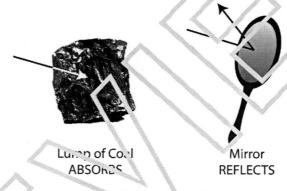

Lump of Coal
ABSORBS

Mirror
REFLECTS

Figure 10.11 Opaque Objects

Transparent materials allow most or all light to pass through them. They are clear, so you can see through them. Light that hits a transparent material is transmitted. It may be slightly refracted, depending on what the material is made of.

Transparent objects can have a color. Think of your sunglasses. You can see clearly through them. They are just a little darker. Sunglass lenses are transparent.

Figure 10.12 Transparent Objects

Translucent materials allow some light to pass through them. They transmit and refract some light, while absorbing or reflecting the rest. We say that translucent materials **diffuse** light. In Figure 10.13, you see a flashlight shining on a thin cloth. Some of the light passes through. But the beam is much wider because some light has been bent (refracted). Also, the beam is not as bright. This is because some of the light has been absorbed by the cloth.

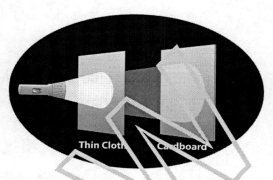

Thin Cloth   Cardboard

Figure 10.13 Translucent Objects

Journal Activity

Objective: Observe what happens to light when it hits the surface of objects.

Notes: List the four things that can happen to light when it hits a surface.

Data: You need five items: a flashlight, a wooden block, a clear or plastic bowl, a small mirror and a glass with water in it. Have your teacher or parent help you collect these items. Turn the flashlight on. Place it on a table facing a wall. Turn the lights off. Place one of the objects on the table between the beam of light and the wall. Record what happened to the light. Draw a picture of it. Do this for each object. Find more objects in your house that you think will absorb, reflect, transmit and refract light. Check your predictions.

Summary: Make four categories for your objects. They should be: absorb, reflect, transmit and refract. Describe how the objects in each category are the same. Describe how the objects in each category are different.

Activity

Experiment using a flashlight and a mirror. Observe how light is affected by the mirror.

CHAPTER 10 REVIEW

The figure below models light shining on a surface of an unknown material. Use the figure to answer the questions.

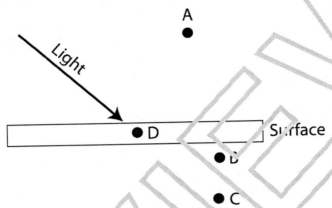

1 If the incoming light is reflected, which point will it strike?

   A   point A   B   point B   C   point C   D   point D

2 If the incoming light is absorbed, which point will it stop at?

   A   point A   B   point B   C   point C   D   point D

3 If the light is refracted, which point will it strike?

   A   point A   B   point B   C   point C   D   point D

4 If the light strikes point A, which type of material could the surface be made of?

   A   wood   C   soil
   B   shiny, polished metal   D   either A or C

5 What does point C represent?

   A   transmission with no refraction
   B   transmission with a lot of refraction
   C   no transmission at all
   D   all reflection and no refraction

Chapter 11
White Light

In the last chapter, we spent a lot of time looking at a flashlight. A flashlight beam is called **white light**. Did you know that white is not really a color? White is all colors put together.

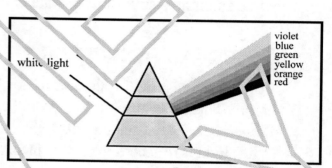

Figure 11.1 Prism

We can see this if we shine white light through a prism. A **prism** is a transparent glass instrument. It can be made of glass or other materials. The prism refracts the white light. As it refracts, the colors of the white light are separated. They separate because some parts of visible light bend more than others as they go through the prism. Look at Figure 11.1. Here again, it is the physical properties of the light and how they behave inside the prism that determines how much the light bends. Not all parts of white light are the same. Some parts of light have more energy than other parts.

Think of it like this—some of your classmates have more energy than others. But you are all part of the same class. The same thing is true for light. All the colors represent the different energy levels. But they all are part of the same light beam.

The parts of light that have the most energy bend the least. That's because they move the fastest through the denser material, the glass prism. They are slowed the least by the glass. Red light bends the least. Violet light bends the most.

Challenge Question
Does violet light have more or less energy than red light? How do you know?

We use the silly name ROY G BIV to describe the colors in white light. Each letter of the name helps you remember one of the colors in white light. The name puts the colors in order from the color most bent in a prism (violet) to the color least bent in a prism (red).

Have you ever seen a **rainbow**? Then you have seen a natural prism at work! When sunlight shines through wet air just right, the water droplets in the air bend light. The white sunlight is split into separate colors.

Meet

Red
Orange
Yellow
Green
Blue
Indigo
Violet

Figure 11.2 ROY G BIV

COLOR AND ABSORPTION

Color is interesting for another reason. Different objects on Earth have different colors. This is because of the way light acts when it hits the surface of the objects. Remember from Chapter 10, light can be absorbed or reflected by a surface. When light is reflected from a surface, it bounces back, right? This means the light can go into our eyes. We can "see" a color. The rest of the light is absorbed by the object. Some colors absorb more light than others.

Lighter colors, like yellow, reflect more light but do not absorb much. White objects reflect all visible light. Darker colors, like indigo, absorb a large amount of light and reflect very little.

Just as white is not a color, black is not really a color either. Black is more like the absence of color. Black objects absorb all visible light and do not reflect any at all.

So, different colors absorb different amounts of light. This can be measured as temperature. **Temperature** is a measure of heat energy. Remember, light is energy. A material that is absorbing light is absorbing energy. That added energy makes the absorbing material warmer.

Dark colored materials will absorb more light and energy than light colored materials. That also means that they will have a higher temperature. That's why black objects left in the Sun get hotter than white objects.

SEEING THROUGH A DIFFERENT LENS

Up to this point, everything we've looked at has been seen with our own eyes. Let's see what **lenses** can do. A lens is a curved transparent object (usually glass or plastic). It forms an image by refracting (bending) light. There are two kinds of lenses: **concave** and **convex**.

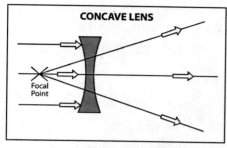

Figure 11.3 Concave Lens

Concave lenses are thinner in the middle than at their edges. Light rays traveling through a concave lens always bend away from each other toward the edges of the lens. After they pass through, it looks like they came from a single point behind the lens. That is a focal point. A concave lens makes images that are smaller than the object.

Convex lenses are thicker in the middle than at their edges. As light waves pass through convex lenses, they bend toward each other. At a certain distance beyond the lens, the light comes together into a point. That point is called the focal point or focus. Convex lenses have many uses. They are used as magnifying glasses and inside microscopes.

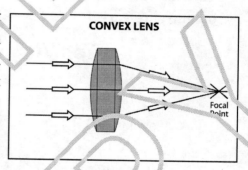

Figure 11.4 Convex Lens

Journal Activity

Objective: To learn about lenses.

Notes: Summarize how light moves through a concave and convex lens. Draw a concave and a convex lens in your journal.

Data. Use the picture here to experiment with different types of lenses. First draw the picture in your journal as it appears below. Then draw how each type of lens alters the picture.

Summary: Summarize how each type of lens changed the picture

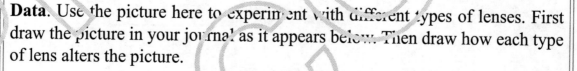

CAMERAS

You've probably used cameras before and wondered how the image you see through the viewfinder ends up on the film you have developed. It all has to do with lenses and mirrors…refraction and reflection.

Cameras come with a convex lens. That lens bends the light so it focuses at a point somewhere behind the lens. When the distance between the lens and film is just right, all the light from each point on the object you want to take a picture of comes together on the film. The film records this pattern of light and *presto chango*…you have a picture!

Some cameras do not put the image directly from the lens onto the film. Sometimes the image is reflected by a mirror behind the lens. The real image is reflected onto a special glass sheet near the top of the camera. When you look through the viewfinder of these kinds of cameras, you are actually using a magnifying glass to see the image better.

Top View of Camera

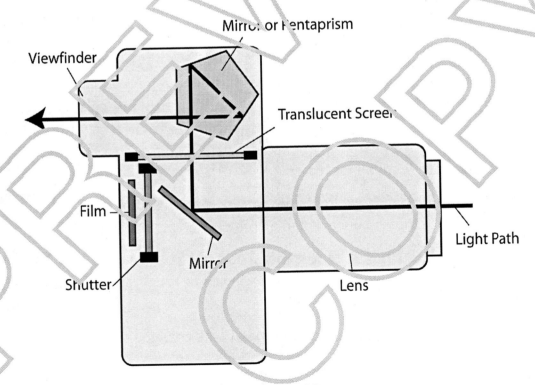

Figure 11.5 Top View of Camera

Journal Activity

Objective: To see how different colors and materials affect the thermal energy absorbed.

Notes: Review how colors and the surface of objects interact with white light waves. Predict what will happen when containers of water are covered with different materials like colored paper or aluminum foil.

Data: Cover several plastic cups with different materials. Use colored paints, construction paper, clear plastic wrap, aluminum foil or cardboard. Then place one cup of water in each covered plastic cup. Put your cups near a sunny window or outside. Measure the temperature of your water at the beginning and every ten minutes for one hour. Then figure out how much the temperature changed during the hour. Record your data in a table.

Summary: What did you notice? Which material or color caused the temperature of the water to increase the most? Which material or color caused the temperature of the water to increase the least? If you were going to a desert, what type of material would you use to make your clothes? Write a short story about your trip and how your clothes helped you.

Activity

Look at the picture of the Apollo Lunar Module. Now look at the picture of the astronaut. What do you notice about the color of the astronaut's suit? What about the Apollo? Do you see the gold foil around the lower part of the module? Why do you think NASA used the gold foil and white suits? Use what you just learned about light waves to help answer these questions

CHAPTER 11 REVIEW

1 **Which of the following colors is not a part of white light?**

    A   yellow   B   red   C   black   D   indigo

2 **Four jars of cold water are placed in a sunny windowsill. Each is painted a different color. Which colored jar will stay the coolest?**

    A   the orange jar   C   the indigo jar
    B   the purple jar   D   the blue jar

3 **What type of instrument would you use to magnify an image?**

    A   convex lens
    B   concave lens
    C   a prism
    D   both a convex and a concave lens

4 **What type of instrument would you use to separate light?**

    A   convex lens
    B   concave lens
    C   a prism
    D   either convex or concave lens

5 **Which color is bent the most in a prism?**

    A   red   B   green   C   blue   D   violet

Chapter 12
Sound

By now, you are familiar with many different sounds. In music class, you may have heard the sound of a piano or a guitar being played. Perhaps you have a played a musical instrument yourself.

Figure 12.1 A Drum

Sound, like light, is a type of energy. Sound energy happens when something vibrates. The physical vibrations of an object cause molecules in the air to get excited. They start bumping into other air molecules close by. When they bump, they transfer their energy or vibration to other air molecules. This transfer of energy makes a sound wave.

This may be hard to imagine because air is thin. We usually don't think it's made up of other things. Remember, it is a gas. Perhaps another example might help. Think of a puddle. Throwing a rock in the middle of the puddle makes waves. The rock makes the water molecules near where it hit move very quickly. They move outward transferring their energy.

Figure 12.2 Puddle

The water molecules vibrate because of the rock hitting the water.

Figure 12.3 Sound Wave

The same thing happens in the air. The vibrations of an object (like the rock) cause the air molecules to squeeze together (compress) and then rapidly spread back out (expand). The result is waves in the air. Regular repeating expansion and compression of air makes **sound waves**. If you are close by, then you can hear sound waves with your ears.

SOUND MOVES THROUGH MATTER

Remember the three states of matter? Solid, liquid and gas. Well as it turns out, sound waves can move through all three states of matter quite easily. Sound is the transferring of vibrating energy. So it only makes sense that you can transfer this sound through any type of matter.

Okay, so you know that sound can move though a **gas**. Remember, you hear sounds as they move through the air. Slamming doors, musical instruments and leaves rustling are all sounds you hear. Air is made up of several gases. Oxygen, hydrogen, helium, nitrogen and water vapor are all parts of air. As we said before, when these molecules vibrate, they create sound waves that travel through the air. Eventually, they reach our ears.

Figure 12.4 A Dolphin Uses Sound Underwater

Sound can move though a **liquid**. Because a liquid is denser than a gas, sound actually moves faster though it. Think about it for a second. The molecules that make up a liquid are packed tighter together. Water is denser than air. They can transfer their energy easier. This makes the sound vibrations travel faster and for longer distances. Have you ever heard a recording of dolphins underwater? Dolphins use sound waves to communicate under water. Try it next time you go swimming! You and a friend get on opposite ends of a pool. Have your friend go underwater and hum. If you go underwater too and listen, you can hear your friend's song!

How does a solid fit in? Do sound waves travel though solids? Of course they do! Sound can move through a **solid**. Haven't you ever heard loud music coming from the room next door? That's sound traveling through the solid wall. How fast do sound waves travel through solids? Remember solids are dense. Their molecules are tightly packed. This makes it easier to transmit the sound wave. So sound must travel faster in solids than in other types of matter.

If you don't believe sound travels fastest through solids, try this experiment. You and a friend stand on opposite sides of a brick or metal wall. Have your friend knock on the wall while you listen with your ear against the wall at the other end. What did you hear? You should be able to hear your friend's knocking, loud and clear!

Challenge Question

Which type of air transmits sound the fastest, dry air or moist air?

Here's the really tough question. Can sound travel through outer space? Let's think about it. Is there any thing in space? Well, there are stars, planets and asteroids. But what about the spaces between these objects? Is there air? Can astronauts breathe in space? NO! They need a space suit to survive because there is no air in space. We say outer space is a vacuum. A **vacuum** is a space that has nothing in it. Remember how a sound wave forms. The molecules of air (or liquid or solid) get excited and bump into each other. If there are no molecules to bump into, the sound cannot be transmitted. So what's the answer to our tough question? Right! **Sound cannot travel through outer space.**

Figure 12.5
The Vacuum of Space

So sound travels the fastest through solids. It travels slowest through gases. It cannot travel through a vacuum.

SOUND AND ITS SPEED

Just like the physical nature of light affects how we see it, the physical nature of sound waves affects how we hear them. The highness or lowness of a sound wave is called **pitch**. It is a measure of how often the air is vibrated. Frequent, fast vibrations make waves that are close together. This makes a high pitched sound. An example is a whistle blowing. Slow vibrations make waves that are spread out. These waves make a low pitch sound. An example is a hound dog barking. Scientists measure pitch using Hertz (Hz).

Low Pitch

High Pitch

Figure 12.6 Sound Waves: High Pitch and Low Pitch

So the object that vibrates determines the pitch of the sound wave. Think of a guitar string. Plucking an open string makes a low pitched noise. However, if you put your finger in the middle of the string, it will make a higher pitched noise. Because the string is half as long, it takes less time to vibrate, so it vibrates faster. This makes the pitch higher. And, what if you increased the length of the guitar string? Exactly! The pitch would be lower.

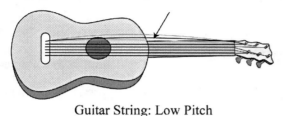

Guitar String: Low Pitch

Figure 12.7 Plucking an Open Spring

Guitar String: High Pitch

Figure 12.8 Plucking Half a Spring

Journal Activity

Objective: To see sound waves.

Notes: Take some notes summarizing sound waves and pitch.

Data: Cover a pie plate with plastic wrap. Place salt grains on the top of the sheet of clear plastic wrap. Sing directly to the plate using a high pitched sound. Sing to the plate using a low pitched sound. Yell at the plate and blow a whistle in its direction. You will see the grains of salt move. Move farther away from the plant and repeat the sounds you made before. Make notes of how the motion of the salt grains changed with the distance from the source.

Summary: Write about your experience. How did the distance from the plate affect the motion of the salt grains?

Journal Activity

Objective: To learn about making sounds.

Notes: Collect an empty rectangular tissue box and some rubber bands. CAREFULLY place the rubber bands around the tissue box. Experiment with different rubber bands to get different sounds. Draw a sketch of your "rubber guitar."

Data: Write a song about sound that you can play on your rubber band guitar.

Summary: Perform your song for your family!

VOLUME

What makes sounds loud or soft? The loudness or softness of a sound is called its **intensity**. It is how much energy is in the wave. Think of a tuning fork. When you strike it, it makes a sound of a certain pitch. If you strike it softly does it make a different sound? No, it makes the same pitched sound. It just makes it quieter. It has less energy because you hit it softly. The arms of the fork don't vibrate the air molecules very much. The force you used to strike the fork with changed, not the length of the fork.

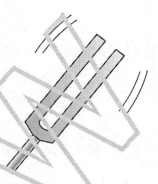

Figure 12.9 Soft Sound

What if you hit the fork harder? What would happen? That's right! It would have more energy and be more intense. It would have a stronger vibration. This would increase the volume of the sound.

How could you get a different pitch from the tuning fork? Right! You would have to change the length of the fork to get a different pitch.

Figure 12.10 Loud Sound

Journal Activity

Objective: To learn about pitch and volume.

Notes: Jot some quick notes about pitch and volume of sound waves. Predict how to make sounds of different volumes.

Data: Using your desk and a ruler, experiment with sounds. Firmly hold the ruler halfway on the desk with the other half hanging off. Using the half hanging off gently, pull the ruler up (don't break it!) and let go. Change the length of the ruler hanging off. Make it longer and shorter. Pull it up a lot or a little.

Summary: Write your observations. What did you notice? How did you make the sound higher pitch? How did you make the sound lower pitch? What caused the sound to change volume?

CHAPTER 12 REVIEW

1 What is sound?

    A   waves created by vibrations of an object

    B   the amount of energy in a sound wave

    C   a space that contains no matter

    D   the loudness or softness of an object

2 What does intensity tell you about a sound wave?

    A   the energy in the sound wave

    B   the speed of the sound wave

    C   the medium that the sound wave traveled through

    D   the amount of matter transferred by the sound wave

3 How can you increase the pitch of a sound made by a tuning fork?

    A   use a longer tuning fork

    B   use a shorter tuning fork

    C   use the same fork but hit it harder

    D   use the same fork but hit it softer

4 What type of sound is made by an object that vibrates slowly?

    A   a low pitch   C   a loud sound

    B   a high pitch   D   a soft sound

5 Which type of rubber band would make the highest pitched sound?

    A   a long rubber band

    B   a medium sized rubber band

    C   a short rubber band

    D   they would all make the same sound

Challenge Question

Which type of musical instrument would make a low pitched sound, a flute or an oboe?

Chapter 13
Work and Simple Machines

Work is moving an object over a distance. A machine has one job: to make work easier. There are **complex machines**, like a crane. These are machines that use motors to work. They are made of many parts. They often have many moving parts. Then, there are the **simple machines**. An **inclined plane** is a simple machine. These machines have no moving parts. There are six types of simple machines. All other machines are made up of a combination of simple machines.

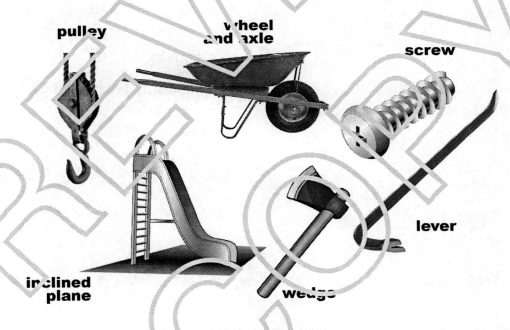

pulley

wheel and axle

screw

inclined plane

lever

wedge

Figure 13.1 Simple Machines

INCLINED PLANE

Inclined planes are also called **ramps**. Inclined planes help us to move from a low place to a high place. They allow us to lift an object in small increments. Have you ever seen a moving van? It has a ramp that rolls out of the back. The ramp allows the movers to push heavy boxes and furniture up the ramp instead of lifting it into the truck. A ramp is one example of an inclined plane. Can you think of another?

The longer the ramp, the easier the object is to move. This is because the incline is less steep. This helps move the object over a longer distance with smaller vertical increases. A long ramp has a smaller angle with the ground. A short ramp has a large angle.

Figure 13.2 Using an Inclined Plane

Thought Activity

Two ramps are placed on either side of a platform. Each has a box placed on it. The boxes are the same

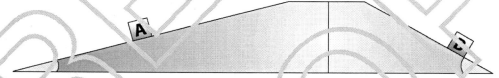

Which box is easiest to move?

Which box has to be moved the greatest distance to get to the platform?

Which ramp is the steepest?

Write 1–2 sentences that describe the difference between the two ramps:

WEDGE

A **wedge** is like two inclined planes stuck together. The wedge is used to push things apart. An example of the wedge is the head of an axe. If you strike at a piece of wood with an axe, the wedge head goes into the wood and splits it open.

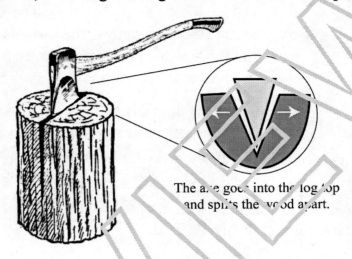

The axe goes into the log top and splits the wood apart.

Figure 13.3 Wedge

A wide axe head works quickly. But you must hit it with more force. A narrow axe head uses less force and works slowly. A chisel and a nail are two more examples of a wedge.

Thought Activity

Two people are splitting wood. Mandy is very strong and likes to get things done fast. Gina is weak but has lots of patience. Match the girl with the wedge she should use.

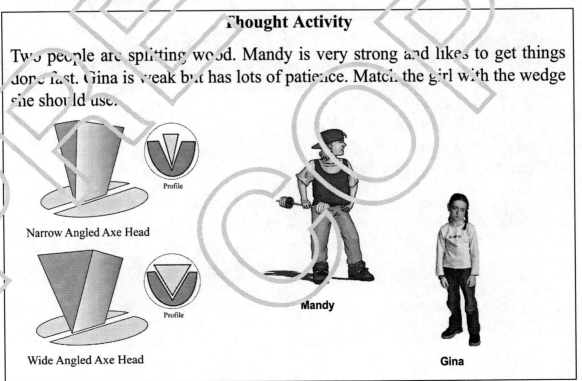

Profile

Narrow Angled Axe Head

Profile

Wide Angled Axe Head

Mandy

Gina

SCREW

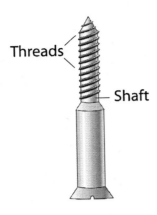

Figure 13.4 A Screw

A **screw** is used to hold things together. A screw can hold wood or metal together. A screw is actually an inclined plane wrapped around a center shaft. The **shaft** is the middle part of the screw. The inclined plane wraps around the shaft to form the **threads**. In the wood screw, the threads cut into the wood at an angle. They create grooves in the wood and hold the screw in place.

When the threads are far apart, the screw is hard to turn. The closer the threads are together, the easier the screw is to turn. But a closely-threaded screw must be turned more times to get it in all the way.

There are many other examples of a screw, like a screw on a bottle top, a corkscrew, a meat grinder and a drill bit.

LEVER

A **lever** is used to move heavy objects. Almost anything can be a lever—even a stick! Laying the lever over a raised point (the **fulcrum**) gives you two arms. These are called the **lifting arm** and the **resistance arm**. Look at Figure 13.5. A man is using a lever to push up a large rock. He pushes down on the lifting arm. The resistance arm lifts the rock.

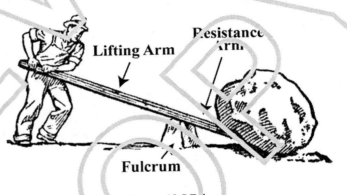

Lifting Arm

Resistance Arm

Fulcrum

Figure 13.5 Fulcrum

Figure 13.6 Hammer Claw

The closer the fulcrum is to the object, the easier the object is to lift. There are many everyday examples of levers. Using a hammer claw to pull a nail out of wood is a lever. Another example is the seesaw on the playground.

PULLEY

Figure 13.7 Piano Pulley

A **pulley** is a wheel with a groove around its middle. A rope fits in to the groove and wraps around the wheel. The wheel can be attached to a surface. The wheel and rope together make up the pulley system. A pulley can be used to easily lift a load. A combination of pulleys can make the load very easy to lift. A weight attached to this pulley system will seem to weigh only 1/2 of its actual weight. A pulley can help move an object into a hard to reach place.

Adding extra pulleys makes it easier to lift something. The more pulleys you add, the easier the work gets. A construction crane often uses pulleys. Another example is on your school's flagpole. The rope and pulley system is used to raise and lower the flag.

Wow! This feels like 50 pounds!

100 pounds

Figure 13.8 A Pulley

WHEEL AND AXEL

The **wheel and axel** is a set of two things that work together to form a machine. A rod or axel in between two wheels is a wheel and axel system. The wheels on the ends of the rod can be large or small. The rod or axel can be long or very short.

Axle

Wheel

Figure 13.9 Wheel and Axel System

Figure 13.10 A Gear

The wheel and axel can work in two different arrangements. If the axel turns the wheel, the wheel carries the weight. Wagon wheels and car tires are two examples. One you might be familiar with is a bike wheel. A bike uses a special wheel and axel called a gear. A **gear** is a wheel with teeth on the outside. The teeth fit together with other teeth in a system. If a wheel turns the axel, the axel carries the weight. An example is lifting a bucket up from a well.

Some more examples of a wheel and axel are a windmill, a roller skate or a rolling pin.

Journal Activity

Objective: To learn about simple machines.

Notes: Take brief notes about the different simple machines.

Data: Go around your house looking for simple machines. Make a table showing the machines you find. Identify each type of machine.

Summary: Write about your experience. Which type of machine did you find most often? Which are most useful?

Journal Activity

Objective: To identify simple machines.

Notes: List the simple machines you have learned.

Data: Draw an example of each type of simple machine.

Summary: Explain how each simple machine helps make work easier. Write a story about someone using each type of simple machine. Make the setting of your story somewhere unusual—like underwater.

Activity

Humans need water to survive. Getting water from a well is one way humans find water sources. Look at the two pictures. One picture shows a modern well and one shows a more ancient design. How can you tell the difference? What are the points A and B representing on the well on the left?

CHAPTER 13 REVIEW

Use the following figure to answer questions 1–2.

 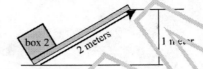

1 **Box 2 is lifted using which simple machine?**

    A   a pulley   C   a lever

    B   an inclined plane   D   a wedge

2 **Box 1 and Box 2 are the same mass. Which box will be easiest to move up to the platform?**

    A   Box 1

    B   Box 2

    C   Both boxes will require the same effort.

    D   Not enough information is given.

Use the following figure to answer question 3.

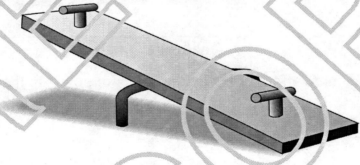

3 **Which simple machine is shown in the figure?**

    A   a pulley   C   a lever

    B   an inclined plane   D   a wedge

4 **Mark sits on one side of the seesaw. Sally stands on the other side. How could you change a seesaw to make it harder for Sally to lift Mark?**

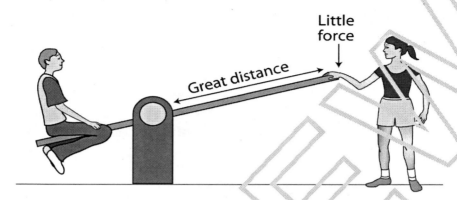

A   Move the fulcrum closer to Sally.
B   Move the fulcrum closer to Mark.
C   Place the fulcrum in the middle of the seesaw.
D   Remove the fulcrum.

5 **A nail can be hammered into a piece of wood, but a bolt cannot. What simple machine can be found at the point of a nail?**

A   a wheel and axel   C   a lever
B   an inclined plane   D   a wedge

Challenge Question

What takes more effort: going up the ladder or going up the slide?

Chapter 14
Force and Motion

WHAT IS FORCE?

A **force** is a push or pull on an object. When a force is **applied** to an object, the motion of the object changes. A force can cause an object to start moving, stop moving or change direction. The change in motion depends on three main factors: direction of force applied, strength of the force and the mass of the object.

DIRECTION OF APPLICATION

The **direction** in which the force is applied determines the direction of motion. The first scene of Figure 14.1 shows a foot that is just about to kick a ball. The force is applied to the ball by the foot. The direction the force is applied is shown by an arrow. The second scene of Figure 14.1 shows what happens after the foot kicks the ball. The ball moves in the direction of the applied force.

Figure 14.1 Kicking a Soccer Ball

Remember that the **angle** at which the force is applied is also important. Figure 14.2 shows that the soccer ball could be kicked in two different angles. You see that the motion that results could be in two different directions. Force A results in Motion A. Force B results in Motion B.

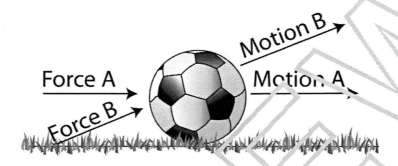

Figure 14.2 Angle of Kicking the Ball

STRENGTH OF FORCE

So, an applied force will move an object in the direction the force was applied in. How *much* the object moves depends on how *much* force is applied. Kicking the ball hard delivers a lot of force, so the ball moves a long way. Kicking the ball gently delivers less force, so the ball does not move far. In Figure 14.3, we can see the aggressive player kicks the ball hard, and it moves a long distance. The lazy player kicks the ball softly, and it only moves a short distance.

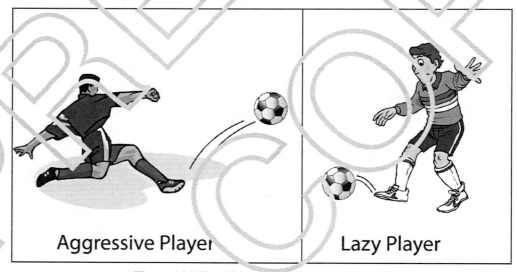

Figure 14.3 Two Players Kick the Ball Differently

MASS OF OBJECT

Mass also affects the motion of the object. The **mass** of an object describes how much matter it contains. What happens if a soccer ball and a bowling ball are each kicked with the same amount of force?

Right, the soccer ball will move much farther than the bowling ball. Why? Because the bowling ball has

Figure 14.4 Mass Affects the Motion

more mass. The greater the mass of an object, the more force must be applied to move the object. So, to move the bowling ball the same distance as the soccer ball, you must use more force on the bowling ball. That is why we use machines to multiply the force that we apply.

For example, using a pulley or lever system we can easily double or triple the force applied to an object. This allows us to move massive objects using a small force.

But what about the force that the object itself can apply? Think about this: If a soccer ball and bowling ball both fall from the sky, which will hit the Earth with more force? That is right, the bowling ball will hit with greater force. That is because the bowling ball has a greater mass.

Figure 14.5 Falling Objects Apply Force

GRAVITATIONAL FORCE

So, we know that objects fall, and that the heavier they are, the more force they strike with. Now, let's look at why things fall in the first place. They fall because the Earth exerts a force on them. This force is called the **gravitational force**. Remember from Chapter 3 that gravitational force is the attraction a massive object has for other objects. The Earth is a lot more massive than a bowling ball. So it exerts a greater attractive force, pulling the bowling ball toward it. The gravitational force that the Earth exerts on all objects is constant. Gravity is what makes things fall when they are dropped. It is also what keeps humans on the Earth's surface and satellites in orbit.

Activity

Consider this situation. A student flicks a marble with force A. What will happen next? Which direction will the marble roll?

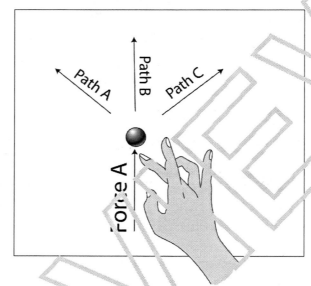

Journal Activity

Objective: To learn about forces applied to objects.

Notes: Review and summarize how mass and forces are related.

Data: Go to the web site: www.bbc.co.uk/schools/scienceclips/ages/6_7/forces_movement.shtml. Do the activities related to force and motion.

Summary: What did you discover? How are the forces needed to move the red truck different from the forces needed to move the blue truck? As a challenge complete the questions and activities at this web site: http://www.bbc.co.uk/schools/ks2bitesize/science/activities/friction.shtml.

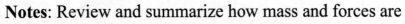

Journal Activity

Objective: To learn about forces applied to objects.

Notes: Review and summarize how mass and forces are related.

Data: Your teacher has supplied you with marbles of different masses. Find the mass of each marble using a balance. Apply the same force to each marble and measure the distance it travels with a ruler. Now try to get all the marbles to travel the same distance.

Summary: Record your efforts in a table. How did the force change between the different marbles? Which marble required the most force? Which marbles required the least force?

Activity

Draw a comic strip showing your favorite superhero kicking a ball during a kickball game. Use at least four frames showing the force applied to the ball and the motion that results.

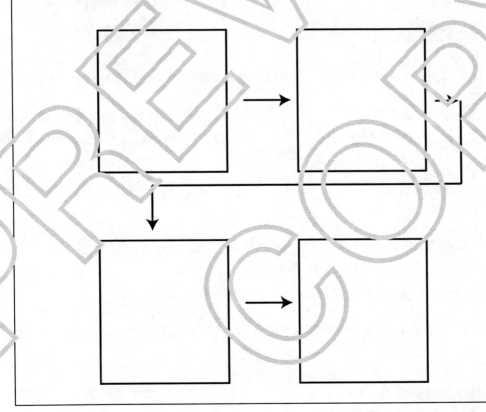

CHAPTER 14 REVIEW

A class puts together the model car ramp shown below. Use the figure to answer questions 1–3.

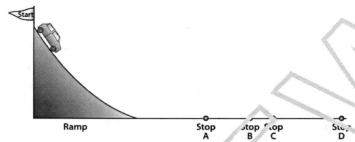

1   Molly's car is placed at the top of the ramp. She releases the car and it rolls to Stop A. What could Molly do to help the car move farther?

    A   Add mass to the car.   C   Clap for the car.

    B   Remove mass from the car.   D   Paint the car.

2   Which student's car will stop at Stop D?

    A   Amy's car (10 grams)   C   Tanisha's car (16 grams)

    B   Serge's car (14 grams)   D   Dora's car (20 grams)

3   Which of the cars from the previous question will stop at Stop B?

    A   Amy's car (10 grams)   C   Tanisha's car (16 grams)

    B   Serge's car (14 grams)   D   Dora's car (20 grams)

4   What BEST describes the direction of Earth's gravitational force?

    A   It pulls objects closer together.

    B   It keeps objects from getting too close.

    C   It pulls objects closer to Earth.

    D   It pulls objects away from the air.

5   Which of the following will have the LEAST effect on the motion of a ball?

    A   the mass of the ball   C   the strength of the force

    B   the direction of the force   D   the weather conditions

Domain Two Review
Physical Science

1   A microscope uses which type of lens?

    A   concave
    B   convex

    C   opaque
    D   absorbing

2   Which type of material is opaque?

    A   glass
    B   plastic wrap

    C   aluminum foil
    D   thin cloth

3   A rainbow is a natural example of a(n)

    A   prism.
    B   concave lens.

    C   convex lens.
    D   opaque material.

4   Sound waves will travel the fastest through which material?

    A   a solid
    B   a liquid

    C   a gas
    D   a vacuum

5   In a piano, when a key is struck a tiny hammer hits a string. Which string below will produce the lowest pitch?

    A   a short string
    B   a medium string

    C   a long string
    D   a very long string

6 How is sound made?

A reflections of an object   C absorption of an object

B vibrations of an object   D refractions of an object

7 Which everyday object below is an example of a lever?

A a slide   C a wishing well

B a steering wheel   D a crowbar

8 What happens when a heavy object that is traveling fast hits a lighter object that is not moving?

A The lighter object goes faster and the heavy object slows down a little.

B The heavier object goes faster and the lighter object stays still.

C The lighter object goes faster and the heavy object also goes faster.

D The heavier object stops completely and the lighter object stays still.

9 What is a force?

A a push or a pull on an object

B a measure of the strength of an object

C it is used to make work easier on an object

D the attraction between two massive objects

10 What keeps humans from flying off the Earth as it spins on its axis?

A the force of gravity

B the atmosphere

C the wind resistance

D the amount of sunlight

Domain Three
Life Science

Chapter 15: Community

S4L1a: Identify the roles of producers, consumers and decomposers in a community.

Chapter 16: Food Web/Food Chain

S4L1b: Demonstrate the flow of energy through a food web/food chain.

Chapter 17: Organisms Adapt to Change

S4L1c: Predict how changes in the environment would affect a community (ecosystem) of organisms.

S4L2a: Identify external features of organisms that allow them to survive and reproduce better than organisms that do not have these features (for example: camouflage, use of hibernation, protection, etc.)

Chapter 18: Changes Affect Organisms

S4L1c: Predict how changes in the environment would affect a community (ecosystem) of organisms.

S4L1d: Predict effects on a population if some of the plants and animals in the community are scarce or if there are too many.

S4L2b: Identify factors that may have led to the extinction of some organisms.

Chapter 15
Communities and Their Organisms

WHAT IS A COMMUNITY?

Where do you live? Do you live in a neighborhood? People often live close to each other in neighborhoods. Sometimes we call a neighborhood a community.

Figure 15.1 A Human Community

Would it surprise you to learn that plants and animals also live in communities? It would?!? Well, it's true. Plants and animals also live in communities. A **community** is all the living things in an area.

Plants and animals are living things. Staying close together helps organisms survive. **Organisms** are living things like plants or animals. Fungi and bacteria are also living things that are part of a community.

Communities are made up of different population of living things. A **population** is a group of the same type of organism existing in an area. Your neighborhood is made up of a population of humans. It is also made up of a population of crows. Or, let's say mice or perhaps grasshoppers. You see, there are hundreds of different populations that make up a single community.

A community is not the same thing as an ecosystem. An **ecosystem** is all the living things in an area *AND* their habitat. Ecosystems are made up of living and non-living things. Plants, animals, water, soil, sunlight and air are all part of an ecosystem.

Figure 15.2
An Ecosystem

In Figure 15.2, the community would be the different bird and plant populations in the picture. As well as the fish, mice, insects and bacteria not seen. While the ecosystem is the birds, plants, water and air in the picture.

WHAT ARE THEIR ROLES?

Figure 15.3 Garbage Men

There are different jobs in a neighborhood. Farmers grow the food. Families eat the food. Garbage men take away the trash. Organisms in a community also have different jobs. Each job, or role, helps the community.

Figure 15.4 Farmer

PRODUCERS

You can think of a farmer like a producer. A farmer makes food for the human neighborhood. **Producers** are living things that make food. A producer makes food for the community. Plants are producers. Grass, trees and bushes are all examples of producers. A producer's role is to make food.

Figure 15.5
Producer

CONSUMERS

Figure 15.6 Family

Families are the consumers of a human neighborhood. They eat the food made by the farmers. **Consumers** are living things in a community that eat food. Animals are consumers. Crows, mice and raccoons are all consumers. A consumer's role is to eat food.

DECOMPOSERS

Decomposers keep the community neat and clean. They clean up all types of waste. Fungi are decomposers. Bacteria and insects are other types of decomposers. A decomposer's role is to clean up waste.

Figure 15.7
Decomposer

Journal Activity

Objective: Learn about a community.

Notes: Summarize the roles of organisms in a community.

Data: Go outside with your teacher or parent. Sit in one area and observe the organisms there. List the different organisms you see. Try to find each type of organism (producer, consumer and decomposer).

Summary: Write about your experience in your own words. Which type of organism was the hardest to find?

Journal Activity

Objective: Learn about different organisms in a community (producer, consumer or decomposer).

Notes: Collect information about a specific type of organism (like grass, caterpillar, mushroom or any other type of living thing you can think of).

Data: Write a creative story from the organism's point of view. Illustrate your story.

Summary: Tell your story to your parents and have them write a quick review.

Journal Activity

Objective: Learn the difference between a community and an ecosystem.

Notes: Summarize the differences between an ecosystem and a community.

Data: Brainstorm different factors (specific things) about communities and ecosystems in a two column table. One column for community and one column for ecosystems.

Summary: Write some rhymes about your different factors. Then combine your rhymes into a creative poem. Write two creative poems, one about communities and one about ecosystems. End each poem with your own definition of a community or ecosystem.

CHAPTER 15 REVIEW

1   What is a community?

   A   a group of organisms living together
   B   an ecosystem
   C   non-living parts of a habitat
   D   living and non-living parts of a habitat

2   What is a decomposer's role in a community?

   A   to make food   C   to clean up waste
   B   to eat food   D   to trap energy

3   What does a maple tree do for a community?

   A   it creates energy   C   it looks nice
   B   it creates food   D   it stores water

4   Which organism below is a consumer?

   A   daisy   C   grasshopper
   B   tulip   D   tomato plant

5   Which organism below is similar to a pine tree?

   A   chipmunk   C   blueberry bush
   B   mountain lion   D   human

Challenge Question

A vulture is a type of bird that eats dead animals. What type of organism is a vulture?

Chapter 16
Organisms and Their Energy

WHERE DO ORGANISMS GET FOOD?

Each type of organism must find food. Food gives living things energy. We can think of food *as* energy. All organisms must get energy from somewhere. Each type of organism get its energy from a certain source. Scientists often group organisms by how they get their energy.

PRODUCERS

Producers make food energy. They get their energy from the Sun. Plants use sunlight, water and soil to get energy. Chemical reactions inside the plant help it make food. They make or "produce" their own food. Plants are producers.

Figure 16.1 Producer

CONSUMERS

Consumers eat organisms for energy. They get their energy from producers or other consumers. They "consume," or eat, their energy. Animals are consumers.

Figure 16.2 Consumers

TYPES OF CONSUMERS

Some animals eat other animals for food. An animal that eats another animal for food is called a **predator**. The animal that gets eaten is called the **prey**.

In Figure 16.3, the moth is prey. The spider is the predator.

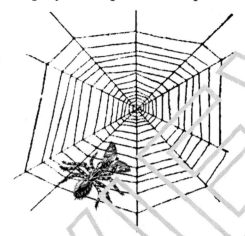

Figure 16.3 Predator and Prey

There are three more specific types of consumers. They are named by the type of food they eat. The three types of consumers are herbivore, carnivore and omnivores. **Herbivores** eat only plants. Deer are herbivores. **Carnivores** only eat other animals. Tigers are carnivores. **Omnivores** eat both plants and other animals. Crows are omnivores. Figure 16.4 shows examples of the different types of consumers.

Herbivore   Carnivore   Omnivore

Figure 16.4 Types of Consumers

Challenge Question

Can you name another example of a herbivore, carnivore and omnivore?

DECOMPOSERS

Decomposers eat dead organisms for energy. They get their energy when other organisms rot. Often, they help break down dead organisms. They "decompose" or break down their energy. They are important in keeping communities clean. Fungi and bacteria are two types of decomposers.

HOW DO ORGANISMS GET ENERGY?

A model called a food chain helps represent energy relationships among organisms. A **food chain** is a picture that shows how each organism gets energy. Links between predator and prey are shown in food chains.

Figure 16.5 A Food Chain

In a food chain, a picture of each organism is separated by an arrow. The arrow points to the organism that gets the energy. In Figure 16.5, the flower gets energy from the Sun. The butterfly gets energy from the flower. The bird gets energy from the butterfly. When organisms die, decomposers use their energy.

Can you spot each type of organism in the food chain? Good! The flower is the producer, butterfly and bird are consumers and the mushroom is the decomposer.

Challenge Question

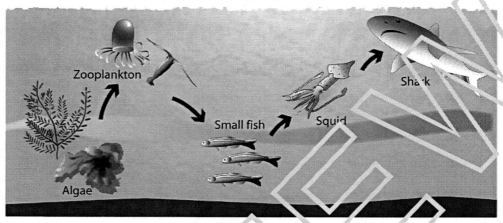

Use the image to answer the following questions.

1. Which organisms are predators?

2. Which organisms are prey?

3. Who is the herbivore?

4. Who are the carnivores?

Journal Activity

Objective: Learn about food chains.

Notes: Summarize what you have learned about food chains.

Data: Draw or make a collage of a food chain.

Summary: Summarize your food chain with words. Use terms like predator, prey, herbivore and carnivore in your summary.

Journal Activity

Objective: Learn about food chains.

Notes: Brainstorm examples of producers and consumers and write them down into two columns.

Data: Put the examples of producers on thin strips of green construction paper and the examples of consumers on thin strips of orange construction paper. Link the strips together with tape or glue like a chain to show their relationships. Don't forget to include the Sun. Draw a sketch of your food chain in the data section. Label each link according to their role.

Summary: Answer the questions below.

1. Food chains contain only one type of which organism?

2. Which type of organism is represented multiple times in the food chain?

Challenge Questions

1. How would decomposers fit into the "food chain" you made above?

2. What would they be linked to?

HOW DO COMMUNITIES GET ENERGY?

One way to show how communities get energy is with a food web. A **food web** is another mode. It shows several different food chains all linked together. A food web is more complicated than a food chain. Food webs show how entire communities get energy.

Remember, organisms live close together in communities. This physical closeness means they can get their energy from several different places. The butterfly in Figure 16.4 travels to hundreds of flowers each day to get energy. It visits flowers from a shrub, grass or a tree.

In addition, the bird doesn't limit its lunch to only butterflies. It also eats moths, beetles, grasshoppers, worms, nuts, seeds or berries.

Lastly, a larger predator like a fox or hawk might eat the bird. The picture that correctly describes their relationship ends up looking more like a web than a chain. In short, plants and animals need each other to survive.

A food web can be very complex. Look at the food web in Figure 16.6.

Figure 16.6 A Food Web

Can you see the individual food chains in the food web? Eventually, all organisms become food for something else. Even the mountain lion will one day be food for decomposers. Also, notice all energy in a community comes from the Sun.

Challenge Activity

Draw two food chains from the food web in Figure 16.6 in the space below.

Journal Activity

Objective: Learn about food webs.

Notes: Summarize what you have learned about food webs.
Data: Draw or make a collage of your own food web.

Summary: Summarize your food web with words. Use terms like predator, prey, herbivore and carnivore in your summary.

Journal Activity

Objective: Learn about food webs.

Notes: Collect information on a particular food web.

Data: Draw a comic strip summarizing your food web starring your own creatively inspired action hero.

Summary: Write a sentence to go with each frame of your comic strip. You can add word bubbles for extra fun. Try to use terms like predator, prey, herbivore and carnivore.

Journal Activity

Objective: Predict how food chains might change.

Notes: Draw a simple food chain with four links.

Data: Consider what would happen to your food chain if one of the links disappeared. Re-draw your food chain with the missing link.

Summary: Write a quick paragraph exploring how the other organisms in your food chain might survive without the missing link. Try to answer these questions. Will other populations get larger or smaller? Will other organisms be able to survive? Are some links (organisms) more important than other links (organisms)?

CHAPTER 16 REVIEW

1   How is a food web different from a food chain?

    A   It is more complex.   C   It is not different.

    B   It is less complex.   D   It is always changing.

2   Herbivores always eat what type of food?

    A   producers   C   omnivores

    B   consumers   D   decomposers

3   A pig eats nuts, berries, fungi and insects. What type of consumer is a pig?

    A   producers   C   omnivores

    B   consumers   D   decomposers

4   Which organism below is a predator?

    A   B   C   D

5   Where do all organisms get their energy?

    A   from other organisms   C   from outer space

    B   from the Sun   D   from dead organisms

Challenge Question

What type of organism is linked directly to all other organisms in a food model?

Chapter 17
Organisms Adapt to Change

WHAT IF THINGS CHANGE?

Have you ever moved? If you have, you might have noticed a new environment can change how you act. You might have to get up earlier to catch the bus. Perhaps you moved to a smaller house. Now you have to share a room with your siblings. Your life has changed because your surroundings changed.

Things on Earth change too. Because the Earth revolves around the Sun, summer becomes winter.

Differences in sunlight affect habitats and their communities. In winter, there is less sunlight. Decreased sunlight means it gets cold in winter. Cold temperatures can change the amount of rainfall in a habitat. Cold temperatures can also affect wind patterns.

Habitat change can influence organism survival. If it gets too cold, animals can't survive. If there isn't enough rain, plants can't survive. Habitat change can affect a whole community. Luckily, most habitat change happens slowly.

Challenge Question

Predict how animals near your home deal with habitat change, like winter. What challenges do they face?

ADAPTATIONS CAN HELP

Habitat shifts affect plants and animals in unique ways. Some trees loose their leaves in the winter. Some animals migrate, or move to a new habitat.

Adaptations for eating fish

Adaptations for catching fish

Figure 17.1 Eagle Adaptations

Both plants and animals have different features they use to survive. We call these features adaptations. An **adaptation** is anything that helps an organism survive. Adaptations take a long time to develop. Change in the appearance of an organism happens over many generations of a population.

For example, eagles have talons to catch fish. How did they get these talons? A long time ago, birds started eating fish to survive. Fish are slippery. Birds needed strong feet to catch and hold onto fish. Birds with the strongest feet caught more fish than birds with weak feet. They lived a long time, raising many children. Their children all had the feature of strong feet. Over time, the adaptation for strong feet, called talons, developed.

But what happened to the fish-eating birds with weak feet? They did not survive. They did not have children. Therefore, their traits were not passed on to future generations. Sometimes adaptations are the way organisms look, or their **external** features. Other times adaptations are the way organisms **behave**.

EXTERNAL ADAPTATIONS

Figure 17.2
Camouflaged Lizard

External adaptations are the physical features of an organism. **Camouflage** is one type of adaptation where animals try to look like their surroundings. Camouflage helps animals to hide. Octopuses change color to look like the ocean floor. Deer and squirrels are brown; they blend in with the forest floor. Turtles often are mistaken for rocks.

Both prey and predators use camouflage. Leopards and tigers have shadowy camouflage. Camouflage helps predators catch more prey. Camouflage is used by many kinds of animals.

Figure 17.3 Camouflaged Predator

Figure 17.4 Warning Coloration

Another type of external adaptation is a **protective** adaptation. Protective adaptations help keep organisms safe. One example is having a hard shell. Beetles have a hard shell. Another protective adaptation is **warning coloration**. Bees, rainforest frogs and venomous snakes are brightly colored. The bright colors warn animals that they are dangerous. Bees can sting and rainforest frogs are poisonous. Venomous snakes bite. The coral snake in Figure 17.4 has warning coloration.

Some animals mimic other things. A **mimic** is something that looks like something else, but really isn't. Stick bugs mimic tree twigs. Some seahorses look like seaweed. Some mimics look like other harmful animals. Hover flies are striped black and yellow like a wasp or bee.

Figure 17.5 A Stick Bug

BEHAVIORAL ADAPTATIONS

Yep, winter's coming. Good night!

Figure 17.6 Hibernation

Behavioral adaptations help organisms survive by changing their actions. When animals hibernate to escape harsh temperatures, they are showing a behavioral adaptation. **Hibernation** is organisms remaining dormant for a long period of time. You can think of it like a special type of sleep. Trees hibernate when they loose their leaves. Chipmunks, bears and frogs also hibernate. Cacti hibernate during dry spells. Many different types of organisms hibernate.

Migration is another behavioral adaptation. Animals **migrate**, or travel, to find a more favorable habitat. Wildebeests, buffalo, whales and birds are just a few examples. Migrations usually happen at the same time each year.

Figure 17.7 Migration

ADAPTATIONS TO SURVIVE WEATHER

Plants and animals deal with habitat extremes in diverse ways. These can be external features, like thick fur, or behavioral features, like hibernation.

Figure 17.8 Frog

In the desert, all organisms must deal with high temperatures. Two types of desert frogs deal with the heat in different ways. One type stays cool by resting in underground tunnels. The other type changes color from brown to white. White reflects more sunlight. Both types of frogs stay cool, but they do so in different ways. One uses an external adaptation while the other uses a behavioral adaptation.

Organisms have adaptations for cold also. Polar bears grow a thick fur coat. They use their coat to stay warm. Pine trees have leaves that do not freeze. This means that the pine tree can keep its leaves all winter. Plants and animals often hibernate to escape weather extremes.

Figure 17.9 Polar Bear

HABITAT CHANGE CAN HELP OR HURT

Sometimes habitat change is helpful. To a pine tree winter is helpful. It keeps its leaves all winter. It can grow while other trees can't. Since other trees have lost their leaves, now there is more sunlight for the pine tree. Sometimes habitat change is harmful. To the other trees, that must loose their leaves, winter is harmful. An organism's adaptations determine if change helps or hurts

Figure 17.10 Changes Can Be Helpful or Harmful

Journal Activity

Objective: Learn how adaptations aid in survival.

Notes: Summarize the different types of adaptations learned so far (camouflage, mimicry, hibernation, protection, migration or warning coloration).

Data: Research different reptiles. Find out ways that some reptiles use camouflage, mimicry, hibernation, protection, warning coloration or migration to survive. Create a table showing the specific reptile and how they use each type of adaptation. Find pictures or make drawings of your reptile.

Summary: Write a sentence about how each reptile uses adaptations.

Journal Activity

Objective: Learn how adaptations help animals survive.

Notes: Perry the Penguin is going on vacation to Africa. Find out information about penguins (what they eat, where they live, how they survive).

Data: Write a story about what Perry needs to pack to survive while in Africa. You can even make a postcard from Perry describing his trip.

Summary: Now write a similar story about your favorite animal.

Journal Activity

Objective: Learn some different types of adaptations in birds.

Notes: Make a word web about the different types of birds in your area. Include the types of food that each type of bird likes to eat.

Data: With your teacher or parent, make or buy different types of bird feeders and place the feeders outside the classroom window. Over a few weeks, observe and record the different types of birds seen at each feeder. Be sure to record information like the date, color of the bird, type of food eaten, and any bird behavior seen. You can even draw a picture of each bird's beak shape. Include hummingbird feeders, several seed feeders, each with a different type of seed, and suet feeders in your study.

Summary: Answer these questions:

Why do you think birds have different types of beaks? Did you notice any similarities in the beaks of seed-eating birds? What about suet-eating birds?

Did you notice any differences in the way different types of birds behaved?

Journal Activity

Objective: Observe some of nature's camouflage.

Notes: All different types of animals use camouflage. Look at the different pictures provided by your teacher.

Data: Predict the type of animal in each picture. Write your prediction in your journal.

Summary: Have your teacher show you where each animal is in the picture. Was your prediction correct? What clues helped you decide?

CHAPTER 17 REVIEW

1   When an organism's body blends in to its environment, what type of adaptation does it have?

    A   hibernation   C   migration
    B   camouflage   D   climate

2   Butterflies need warm temperatures to fly. What type of change would be helpful to butterflies?

    A   nighttime   C   summer
    B   winter   D   autumn

3   When winter comes to the arctic, temperature can drop below –40°C. How do Canadian geese adapt to this change?

    A   They camouflage.   C   They migrate.
    B   They hibernate.   D   They die.

4   What type of adaptation would help an animal survive in the cold?

    A   long legs   A   bright colors
    B   a thick fat layer   B   sparse feathers

5   Arctic hares are white in the winter and brown in the summer. What type of external adaptation does it have?

    A   mimicry   C   migration
    B   hibernation   D   camouflage

Challenge Questions

Can you think of an adaptation plants or animals might use to survive a hot, dry summer?

How does looking like a wasp help a hover fly?

Chapter 18
Changes Affect Organisms

WHAT IF THE CHANGE IS TOO QUICK?

FAST HABITAT CHANGE CAN HURT

Humans destroy natural habitats for houses, farms and shops. In about a day, a lush forest can be completely destroyed. The forest is transformed into something completely different. Many animals and plants are displaced by habitat destruction. Habitat destruction is one change that can happen quickly. Another

Figure 18.1 Air Pollution

fast habitat change is drought. In 2007, a drought struck Georgia. This caused many organisms harm. Plants didn't grow. Animals searched for water. Fast habitat change can be harmful.

Pollution is another example. Humans can release lots of it. This can kill organisms. Soot covers tree leaves. Smog blocks sunlight. Chemicals poison animals.

Giant panda bears were displaced by humans. Habitat destruction and pollution threatened the pandas. In the 1980s, pandas were placed on the endangered species list. **Endangered** means very few organisms of that type are alive. The endangered species list was created in 1963. An international group called The International Union for the Conservation of Nature (IUCN)

Figure 18.2 Panda

maintains the list today. This list changes each day. Efforts to help the panda have been mostly successful.

EXTINCTION

Endangered organisms can become extinct. **Extinction** occurs when all organisms of one type die. Passenger pigeons are extinct. Both plants and animals can become extinct. Often, more than one factor causes organisms to go extinct.

Here's an example of an extinct organism:

Woolly mammoths were fur covered elephant-like animals. They lived during the ice age. They went extinct around 8,000 years ago. Scientists know that around 8,000 years ago the Earth's climate was getting warmer. Humans and other predators hunted mammoths. So, possible reasons for mammoth extinction are:

Figure 18.3 Mammoth

- climate change.
- human hunting.
- too many predators.

Can you think of a fourth factor that might have made woolly mammoths extinct? Notice, many answers will fit here. Things like no food, no water, or diseases may have made mammoths go extinct. Any of these answers are correct.

EXTINCTION STILL HAPPENS TODAY

Figure 18.4 Tasmanian Tiger

Extinction doesn't only happen to woolly mammoths and dinosaurs. Extinctions can happen at any time. Any plant or animal can go extinct. The dodo bird is one example of an animal that went extinct during human history. Some animals went extinct so recently, they can be seen in pictures! The Tasmanian tiger is one example. Figure 18.4 shows a picture of a Tasmanian Tiger. Just like the endangered species list, the list of extinct animals changes each day.

Fast Habitat Change Can Help

Figure 18.5
Bacteria Eating Pollution

To some, quick habitat change is helpful. Some bacteria like to eat pollution. When it is released, there is a lot of food. To bacteria, pollution is a good thing.

Beavers make ponds. They use their new habitat to live. Beavers and fish benefit when a new pond is made. However, field mice are harmed. Sometimes changes benefit some while harming others. Again, it depends on an organism's adaptations.

Figure 18.6 Beaver Pond

Population Controls

Recall communities exchange energy through food webs. Usually, communities are balanced. Predators keep the amount of prey low. Limited prey keeps the number of predators down. Each one controls the other.

Figure 18.7 Drought

Now think about how habitat change can affect organisms. Disease or extreme weather can also affect the level of organisms. Drought causes plants to decrease. This affects the entire community. A **scarcity** or lack of plant life will harm other organisms. If there aren't enough plants, animals starve. When there is a scarcity of organisms, we can say it is **underpopulated**.

Too many organisms in an area mean it is **overpopulated**. This can also be harmful. Too many deer in an area will eat ALL the plants. This is harmful to the plants, rabbits, chipmunks, squirrels, finches, turtles and other deer. Lots of deer can lead to a scarcity of plants. This affects the entire community.

All communities are connected. A scarcity of one organism might result in an overpopulation of another. If there are fewer oak trees in a forest, there is more room for maple trees.

On the other hand, a scarcity of one organism might result in a scarcity of another organism. Fewer insects mean fewer insect-eating birds. It all depends on the community and its interactions.

Where did all the insects go? I'm starving!

Figure 18.8
Bird Eating Insect

Journal Activity

Objective: Learn how factors cause extinctions in organisms.

Notes: Below is a list of animals that have gone extinct. Select one animal, and do some research to find out when the animal went extinct and why.

Extinct Animals:

Allosaurus, Passenger Pigeon, Great Auk, Carolina Parakeet, Western Black Rhinoceros, Sea mink, Golden Toad, Galapagos saddle-backed tortoise, Quagga, Tasmanian Wolf, Bali tiger and Japanese sea lion.

Data: Make a word web showing at least four factors that might have caused the extinction of your animal.

Summary: Write a creatively inspired non-fiction story about how your animal might have avoided extinction.

Journal Activity

Objective: Predict what will happen to different populations of organisms when the abundance of other organisms in the food web changes.

Notes: Draw a food web with at least four different types of organisms.

Data: Write a quick prediction summarizing how increasing or decreasing the different populations will affect the other organisms in the food web.

Summary: Discuss your predictions with your classmates and teacher. Draw a new food web showing your prediction. Make a bar graph before and after the population changes showing how the numbers of organisms in your food web changed.

Journal Activity

Objective: Predict the effects of a flood on a community.

Notes: Every so often, a river in the United States floods. Silt from the river fertilizes soils. Underwater grasses provide temporary cover for baby fish, fish eggs, insects and turtles. Trees that get uprooted and fall into the river provide new habitat for many animals in and near the river.

Data: Collect information about the types of plants and animals in and around rivers near your home. Be sure you include organisms that live on land around the river. Birds, reptiles, mammals, fish and insects should all be included.

Summary: Predict how flood would impact the animals in your river. Who is helped? Who is harmed?

CHAPTER 18 REVIEW

1 **When there are too many of one type of organism, we say they are**

    A   scarce.   C   overpopulated.

    B   underpopulated.   D   rare.

2 **Which word below might also describe endangered organisms?**

    A   overpopulated   C   numerous

    B   scarce   D   plentiful

3 **What did NOT cause the dinosaurs to go extinct?**

    A   humans   C   disease

    B   lack of food   D   changing climate

4 **How would a scarcity of insects affect a population of the animal seen here?**

    A   It would make the frog population larger.

    B   It would make the frog population smaller.

    C   It would keep the frog population at the same size.

    D   It would have no impact on the frog population.

5 **What type of habitat change would hurt a penguin?**

    A   getting warmer   C   more fish

    B   getting colder   D   rising sea levels

Domain Three Review
Life Science

1 **What is the role of a corn snake in a community?**

   A   It is a producer.

   B   It is a consumer.

   C   It is a decomposer.

   D   It has no role in a community.

2 **Which organism type below is responsible for keeping communities neat and clean?**

   A   producer   C   decomposer

   B   consumer   D   adaptation

3 **Where do carrots get their energy?**

   A   from the Sun   C   from deer

   B   from other carrots   D   from humans

4 **Skunks are black and white. They are easily seen in their habitat. How does this coloration help the skunk?**

   A   It warns other animals to stay away.

   B   It invites other animals to come closer.

   C   It is camouflage for the skunk.

   D   It helps the skunk capture more prey.

5 **The Northern Flicker is a woodpecker. Every winter this bird flies south. How would you classify this adaptation?**

   A   external   C   behavioral

   B   protective   D   mimicry

6 **What does the picture below represent?**

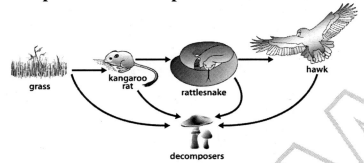

grass   kangaroo rat   rattlesnake   hawk

decomposers

A a food chain	C an ecosystem
B a food web	D a biome

7 **Warmer temperatures make the Summer Sizzle flower population increase. Maiden Lark beetles eat only Summer Sizzle flowers. How will warmer temperatures affect the population of Maiden Lark beetles?**

A It will increase.	C It will stay the same.
B It will decrease.	D It will be scarce.

8 **Great white sharks are grey on top and white on bottom. This helps the shark hide from prey while hunting. What type of adaptation is the color shading in sharks?**

A protective	C mimicry
B camouflage	D warning coloration

9 **Algae in a stream use sunlight to grow. Crayfish eat algae. Raccoons eat crayfish. What do you think will happen to the raccoon population if lots of trees grow along the stream bank?**

A It will increase.	C It will stay the same.
B It will decrease.	D It will be overpopulated.

10 **Seahorses look like floating seaweed. How does this help the seahorse?**

A   It helps the seahorse play games with other animals.

B   It helps the seahorse hide from predators.

C   It helps the seahorse stay warm.

D   It helps the seahorse look cool.

Georgia 4th Grade CRCT
Post Test 1

1 Martin used a ramp to load his dirt bike in the back of his pick up truck. What type of simple machine did he use?   S4P3a

   A   screw

   B   pulley

   C   wheel and axel

   D   inclined plane

2 What type of consumer is a rabbit?   S4L1b

   A   herbivore

   B   carnivore

   C   omnivore

   D   decomposer

3 A scientist wants to collect images of objects in the outer parts of the solar system. Which instrument should he use?   S4E1d

   A   a barometer

   B   binoculars

   C   a space shuttle

   D   a probe

4 What is a consumer's role in a community?   S4L1a

   A   to eat food

   B   to eat wastes

   C   to make food

   D   to hibernate

5 The four planets closest to the Sun are called   S4E2d

   A   Jovian planets.

   B   Gas Giants.

   C   terrestrial planets.

   D   dwarf planets.

6   Which pipe in the pipe organ will make the lowest pitched sound?   S4P2b

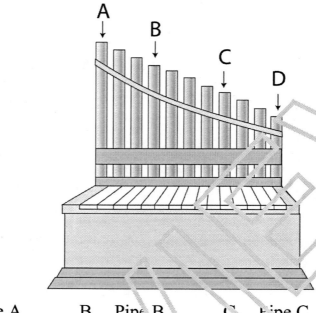

A   Pipe A   B   Pipe B   C   Pipe C   D   Pipe D

7   The boiling point of water is 100°C. At this temperature, the water changes   S4E3b

A   from a solid to a liquid.
B   from a solid to a gas.
C   from a liquid to a gas.
D   from a liquid to a solid.

8   When carrying a box up an inclined plane, what force are you trying to overcome?   S4P3d

A   the force of air resistance   C   the friction of your feet
B   the force of gravity   D   the heat of motion

9 You were doing an experiment to figure out how many
consumers you could keep in an aquarium. Which organism
below would NOT be appropriate to use in your experiment?

S4L1a

A

C

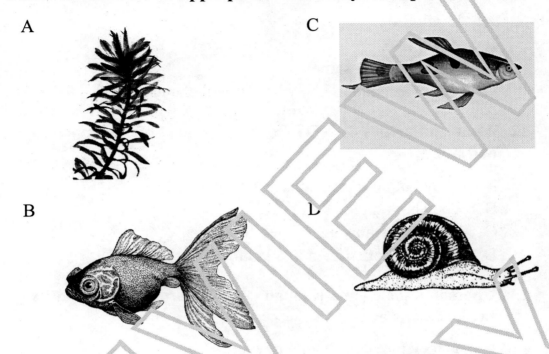

B

D

Use the following figure to answer questions 10 and 11.

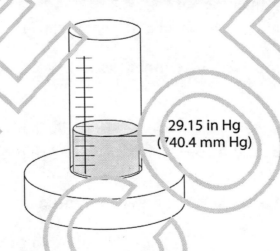

29.15 in Hg
(740.4 mm Hg)

10 What weather instrument is shown in the picture?

S4E4a

A anemometer

B thermometer

C weather vane

D barometer

11 **The forecast for tomorrow says the weather will improve. It will be sunny and dry. How will the barometer change between now and tomorrow?**   S4E4a

A   It will drop.
B   It will rise.
C   It will stay the same.
D   It's impossible to know how it will change

12 **What similarities do NOT exist between a food chain and a food web?**   S4L1b

A   They both use the Sun as an energy source.
B   They both include producers
C   They both show energy relationships.
D   They are both drawn in a straight line.

13 **A person knocks on a wooden door. Where do sound waves travel the slowest?**   S4P2a

A   as they move through the wood
B   as they move through the air
C   as they enter the ear
D   all sound waves travel at the same speed

14 **What astronomical instrument can travel the farthest in space?**   S4E1d

A   a telescope   C   a binoculars
B   a space shuttle   D   a probe

15 **What force is constant everywhere on Earth?**   S4P3d

A   friction   C   air resistance
B   gravity   D   surface area

16 Seasonal rains in Africa supply much of the water grasses
    need to grow. As a result, many grasses only grow during the
    wet season. Different regions in Africa have rain during different
    times of the year. How do wildebeests (herbivores) deal with this
    fluctuation in food supply?   S4L2a

    A   They hibernate.
    B   They migrate.
    C   They develop camouflage.
    D   They do nothing. Wildebeests eat insects.

17 Water exists naturally on Earth as   S4E3a

    A   a solid.
    B   a liquid.
    C   a gas.
    D   a solid, liquid and gas.

18 What type of adaptation is NOT helpful to a duck?   S4L1a

    A   waterproof feathers   C   large talons
    B   webbed feet   D   sensitive bill

19 What happens when light is refracted?   S4P1b

    A   It goes straight through a surface.
    B   It gets absorbed by a surface.
    C   It gets bent as it goes through a surface.
    D   It bounces back.

20 Earl records seasonal changes in his hometown. He uses
    monthly averages of temperature and precipitation to do this.
    What is he observing?   S4E4d

    A   the weather
    B   the climate
    C   the air mass
    D   the water cycle

21 A raccoon eats crayfish and small fish in a creek. It also eats nuts and berries collected along the bank. What type of consumer is a raccoon? S4L1b

A herbivore C omnivore
B carnivore D decomposer

22 Constellations appear to move across the sky from month to month because S4E1

A the Earth revolves around the Sun.
B the stars revolve around the Earth.
C the Sun revolves around the Earth.
D the stars move faster than the planets.

23 Which of the following verbs BEST describes the production of sound? S4P1a

A swinging C singing
B vibrating D varnishing

24 Humans drain a swamp to build stores. Which organism is MOST negatively impacted? S4L1c

A humans C fish
B raccoons D crows

25 Which of the following means "decreasing" when talking about the phases of the Moon? S4E2b

A waning C proceeding
B waxing D diminishing

26 A water buffalo uses its horns to fight off attacking predators. What type of adaptation are horns to water buffalo? S4L2a

A protective C hibernation
B mimicry D warning coloration

27 **Why do black objects get the hottest when placed in sunlight?** S4P1a

   A   They absorb more light energy.
   B   They absorb less light energy.
   C   They reflect more light energy.
   D   They refract less light energy.

28 **Different areas of pressure are shown on a weather map using** S4E4b

   A   lines with triangles.
   B   shaded areas.
   C   an "H" and an "L."
   D   lines with semicircles.

29 **What does the picture below show?** S4L1b

   A   a food web   C   a life cycle
   B   a food chain   D   a population

30 **Which of the following statements about the Sun is TRUE?** S4E1a

   A   The Sun is the biggest and brightest star in our galaxy.
   B   The Sun is the closest star to Earth.
   C   The Sun is located outside the Milky Way galaxy.
   D   The Sun's distance from Earth is measured in light years.

31 What type of lens is thin in the middle? S4P1c

A   convex   C   prism
B   concave   D   kaleidoscope

32 How can planets be distinguished from the stars in the night sky? S4P1b

A   Planets always appear bigger than stars.
B   Stars and planets are different colors.
C   Planets shine with a more steady light than stars.
D   Stars appear to move while planets remain in the same positions.

33 Which color beach umbrella stays the coolest? S4P1a

A   black   B   white   C   red   D   blue

34 Which of the following is NOT a form of precipitation? S4E3e

A   fog   B   hail   C   snow   D   sleet

35 How would a scarcity of lions affect the animal pictured here? S4L1d

A   Its population would increase.
B   Its population would decrease.
C   Its population would stay the same.
D   Its population would become endangered.

36 We experience day and night because   S4E2a

  A   the Earth rotates on its axis.

  B   the Earth revolves on its axis.

  C   the Earth revolves around the Sun.

  D   the Earth rotates around the Sun.

37 How would an abundance of *Eucalyptus* trees affect the   S4E1d
  animal pictured here?

  A   Its population would increase.

  B   Its population would decrease.

  C   Its population would stay the same.

  D   Its population would move to a new habitat.

38 Earth's axis is tilted 23.5°. If it were not tilted, what would   S4E2c
  happen?

  A   Seasons would be shorter.

  B   Seasons would be longer.

  C   There would be no day and night.

  D   There would be no seasons.

39 Which object will get the hottest when left in sunlight?   S4P1a

  A   a white cardboard box   C   a lump of coal

  B   a mirror   D   a yellow flower

40 The color of the hottest star is S4E1a

A red.

C yellow-white.

B blue.

D blue-white.

41 When do ice crystals form in clouds? S4?3c

A when clouds form at low altitudes

B when clouds form in warm air

C when clouds form in very cold air

D when clouds produce precipitation

42 The path of a beam of light is represented by the straight line in the diagram below. What is happening to the beam of light? S4P1b

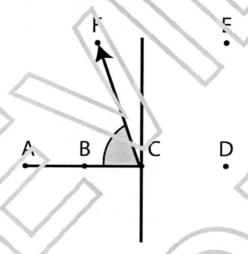

A It is being absorbed.

B It is being transmitted.

C It is being refracted.

D It is being reflected.

43 The water cycle circulates fresh water S4E3a

A within the surface of the Earth.

B between the Earth and the atmosphere.

C between the oceans and lakes.

D between the atmosphere and space.

44 Which plant below would be MOST harmed by a very dry summer?   S4L1c

     A   pine tree   C   oak tree

     B   water lily   D   blackberry bush

Use the following figure to answer questions 45 and 46.

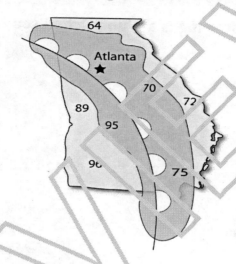

45 What is the weather like in Atlanta?   S4E4b

     A   around 45°F and clear   C   around 70°F and clear

     B   around 69°F and rainy   D   around 90°F and rainy

46 What will the weather be like in Atlanta in 12 hours?   S4E4b

     A   around 90°F and drier   C   around 65°F and drier

     B   around 75°F and wetter   D   around 90°F and wetter

47 Wild pigs eat many things. One of their favorite foods is acorns. When pigs eat the acorns they do not grow into oak trees. Which area do you expect would have the smallest pig population?   S4L1c

     A   an area with lots of baby oak trees

     B   an area with few baby oak trees

     C   an area with no baby oak trees

     D   an area with only baby maple trees

48 As the Moon proceeds from new to full moon, the lighted part of the Moon   S4E2b

   A   increases.

   B   decreases.

   C   increases then decreases.

   D   stays the same.

49 Which of the following is a high pitched sound?   S4P2b

   A   a baby's cry

   B   a dog's growl

   C   a car engine

   D   a drum

50 What main idea do food chains and food webs show?   S4L1b

   A   energy movement   C   populations

   B   lifecycle changes   D   organism adaptations

51 Any cloud that touches the ground becomes   S4E3e

   A   a stratus cloud.   C   ice.

   B   precipitation.   D   fog.

52 Where can sound NOT travel?   S4P2a

   A   through a gas   C   through a solid

   B   through a liquid   D   through a vacuum

53 As the Earth revolves around the Sun, the Sun's rays   S4E2c

   A   strike the surface of the Earth in the same way.

   B   always strike the surface of the Earth at the poles.

   C   strike the surface of the Earth differently.

   D   always strike the surface of the Earth at the equator.

54 Cane toads were introduced in 1935 in Australia. They have no natural predators in Australia. They can eat a wide variety of foods. By 2005, which word MOST LIKELY describes the cane toad population in Australia?   S4L1c

A scarce

B under-populated

C endangered

D over-populated

55 Which of the following statements about constellations is FALSE?   S4E1c

A Constellations help us remember areas of the sky.

B Constellations are visible for a few months at a time.

C The movement of constellations is predictable.

D The shape of constellations changes every few months.

56 An abundance of which organism below would help a bald eagle?   S4L1d

A mice   B fish   C owls   D hawk

57 Which will bend light the most?   S4P1a

A air

B water

C glass prism

D lump of coal

58 Above 100° Celsius, water is   S4E3b

A a liquid.

B snow and ice.

C a solid.

D a gas.

59 What do foxes do in a community?   S4L1a

A make food

B consume food

C break down wastes

D hide other organisms

60 Look at the figure. Two people are playing billiards. A player wants to hit ball 2 with ball 1. Where should he strike ball 1? S4P3b

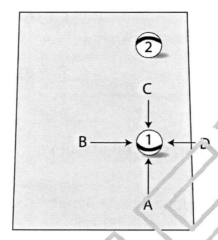

A   Point A
B   Point B

C   Point C
D   Point D

61 What type of simple machine is attached at point A in the wishing well pictured here? S4P3a

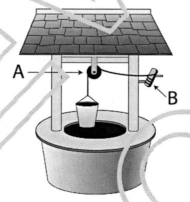

A   screw
B   pulley

C   wheel and axel
D   inclined plane

62 Where is the BEST place for a weather vane? S4E4a

A   on a rooftop that is exposed to the wind
B   on the ground under trees and other objects
C   on a branch of a really tall tree
D   inside next to an open window

63 A steel ball has a mass of 6 grams. A plastic ball has a mass of S4P3b
1 gram. If both balls are the same size, which ball will hit the
floor with a greater force when dropped?

   A   the steel ball

   B   the plastic ball

   C   both balls

   D   neither ball

64 Which organism could replace the shark in the food web? S4L1b

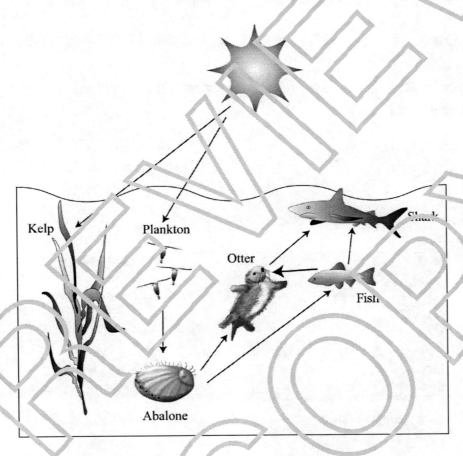

   A   killer whale   C   salmon

   B   shrimp   D   seal

65 Which type of drum will produce a low pitched sound?   S4P2b

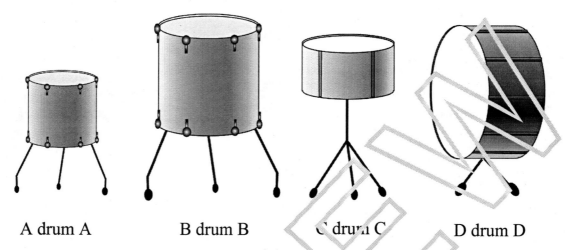

A drum A   B drum B   C drum C   D drum D

66 Frozen water melts into liquid water when it reaches what temperature?   S4E3b

A   0º Celsius   C   50º Celsius
B   32º Celsius   D   100º Celsius

67 Which organism below is a decomposer?   S4L1a

A

C

B

D

68 **Which wind instrument will make the lowest pitched sound?** S4P2b

Flute

Oboe

Saxophone

Clarinet

A flute
B oboe

C saxophone
D clarinet

69 **Four people are playing basketball. The player with the ball** S4P3c
wants to pass the ball to her teammate. Which force should
she apply to the ball to make it go to her teammate?

A Force A
B Force B

C Force C
D Force D

70 **The hockey player strikes the puck at force 1. Which direction will the puck travel?**

S4P3b,c

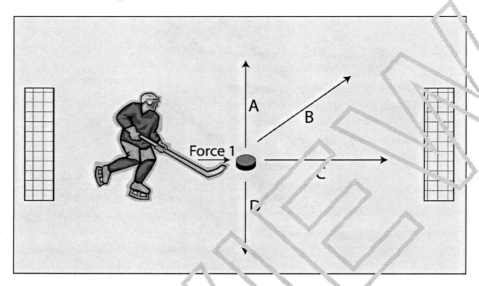

A   Direction A

B   Direction B

C   Direction C

D   Direction D

Georgia 4th Grade CRCT
Post Test 2

Use the following figure to answer questions 1 and 2.

1   What type of weather instrument is shown in the picture?   S4E4a

     A   barometer

     B   anemometer

     C   rain gauge

     D   wind vane

2   Weather forecasters use this instrument to   S4E4a

     A   measure changes in air pressure.

     B   measure wind direction.

     C   measure wind speed.

     D   measure the amount of rainfall.

3 **How would you BEST summarize the difference between a consumer and a producer?** S4L1b

    A   Producers are small, and consumers are large.

    B   Producers make food, and consumers eat food.

    C   Consumers make food, and producers eat food.

    D   Producers are green, and consumers are multicolored.

4 **Intensity is another way to describe a sound's** S4P2a

    A   duration.   C   volume.

    B   length.   D   pitch.

5 **The Sun heats the ground and causes warm air to rise. This process is called** S4E3c

    A   condensation.   C   convection.

    B   conduction.   D   precipitation.

6 **Brass is a metal that is denser than steel. Salt water is denser than fresh water. Pick the situation below where sound will travel the fastest.** S4P2a

    A   a person knocking on a brass pipe under the ocean

    B   a person knocking on a steel pipe under the ocean

    C   a person knocking on a brass pipe under a lake

    D   a person knocking on a steel pipe under a lake

7 **What is a daisy's role in a community?** S4L1a

    A   to eat food   C   to make food

    B   to eat wastes   D   to hibernate

8 **What is the coolest star color?** S4E1a

    A   red   B   blue   C   white   D   orange

9   **Which organism below is NOT a producer?**   S4L1a

A

C

B

D

10 **Edmund observes the stars with his eyes. Then he observes**   S4L1d
     them using a telescope. How does the telescope change what
     he sees?

  A   He sees more stars than he can with his eyes.
  B   He sees fewer stars than he can with his eyes.
  C   The stars look smaller and less bright with the telescope.
  D   He cannot see any stars with the telescope.

11 **Which color of white light contains the MOST energy?**   S4P1a

  A   red   B   yellow   C   green   D   black

12 **What does the arrow in a food chain represent?**   S4L1b

  A   energy movement
  B   lifecycle changes
  C   growth of organism
  D   organism adaptations

13 Which statement BEST describes the movement of the planets?   S4E1c

    A   They move in unpredictable ways around the Sun.
    B   They move together around the Sun.
    C   They move at different speeds around the Sun.
    D   They move at the same speed around the Sun.

14 What type of weather occurs where there is high pressure?   S4E4a

    A   cloudy and rainy   C   cloudy and windy
    B   cloudy and stormy   D   sunny and dry

15 Which thing below do producers NOT use to get energy?   S4L1b

    A   sunlight   C   water
    B   soil   D   consumers

16 A trombone makes different notes by changing the positions of a slide. How should a musician position the trombone slide to make a high pitched sound?   S4P2b

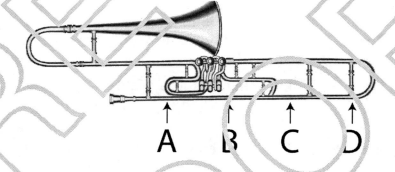

    A   position A   C   position C
    B   position B   D   position D

17 The Northern Hemisphere is tilted away from the Sun. What season is it in Georgia?   S4E2a

    A   summer   C   spring
    B   winter   D   fall

18 According to the food web shown here, which organisms get their food from more than one source? S4L1b

    A   deer, flower, tree and insects

    B   flower, deer, snake and bird

    C   mushroom, insects, snake, bird, deer and hawk

    D   mushroom, insects, mouse, snake and tree

19 Which of the following correctly lists the first four planets in order from closest to the Sun to farthest from the Sun? S4E2d

    A   Venus, Mercury, Earth, Mars

    B   Mars, Venus, Earth, Mercury

    C   Mercury, Venus, Earth, Mars

    D   Venus, Earth, Mars, Jupiter

20 Steven is skipping rocks at a pond one day. He notices that when he throws a rock it makes a curved path as it falls into the pond. What force causes the rock to have a curved path when thrown? S4P3d

    A   wind   C   gravity

    B   rain   D   electricity

21 **A bat captures and eats moths. What type of consumer is a bat?** S4L1b

    A   herbivore   C   omnivore

    B   carnivore   D   decomposer

22 **What cloud is highest in the sky and made of ice crystals?** S4E3c, e

    A   cumulus clouds   C   cumulonimbus clouds

    B   stratus clouds   D   cirrus clouds

23 **A recorder makes sound by opening holes along its shaft. When a hole is opened, air escapes making a sound. This is how the length of the recorder is changed to make sounds of different pitches. Which hole should Vanessa open to make a high pitched sound?** S4P2b

    A   hole A   C   hole C

    B   hole B   D   hole D

24 **Which of the following is MOST responsible for creating the seasons?** S4E2c

    A   Earth's rotation on its axis

    B   the tilt of the Earth's axis

    C   Earth's revolution around the Sun

    D   the rate of the Earth's rotation

25 **What force causes all objects on Earth to fall when dropped?** S4P3d

    A   gravity   C   air resistance

    B   electricity   D   friction

26 **The path of a light beam is represented by a straight line in the diagram below. The diagram shows the light beam being** S4P1a

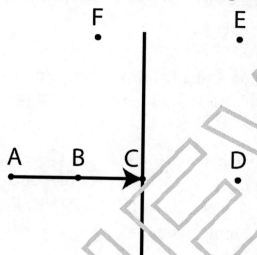

A   absorbed.

B   transmitted.

C   refracted.

D   reflected.

27 **The stars in the night sky are** S4E1a

A   different sizes and brightness.

B   the same size and brightness.

C   the same distance from Earth.

D   really close to Earth.

28 **What would happen to the consumers in an area if all the producers died?** S4L1d

A   Their population would increase.

B   Their population would decrease.

C   Their population would stay the same.

D   Their population would become overpopulated.

29 **A rising barometer means** S4E4a

A   the weather is getting worse.

B   the weather is improving.

C   the weather will not change much.

D   the wind speed is increasing.

30 How would you classify an earthworm in a community?   S4L1a

   A   as a producer   C   as a decomposer

   B   as a consumer   D   as a predator

31 Erica is going on a picnic this evening. She wants to know if   S4E4a
the weather will get better or worse. What instrument will tell
her this?

   A   weather vane   C   anemometer

   B   thermometer   D   barometer

32 What type of simple machine is shown at point B in the   S4P3a
diagram?

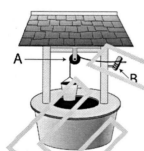

   A   screw

   B   pulley

   C   wheel and axel

   D   inclined plane

Use the following figure of a cumulonimbus cloud to answer question 33.

33 What form of precipitation does this type of cloud generally   S4E3e
produce?

   A   snow   B   fog   C   sleet   D   rain

34 **What type of lens do scientists use to study really small objects?** S4P1c

    A   convex   C   prism

    B   concave   D   kaleidoscope

35 **When the Moon is fully lit, what phase is it in?** S4E2b

    A   first quarter   C   new moon

    B   full moon   D   third quarter

36 **Which object will reflect light the BEST?** S4P1b

    A   aluminum foil

    B   black construction paper

    C   a piece of wood

    D   a blue tee shirt

37 **What do snakes do in a community?** S4L1

    A   make food   C   break down wastes

    B   consume food   D   hide other organisms

38 **How do objects in space appear through a telescope?** S4E1d

    A   They appear smaller.

    B   They appear larger.

    C   They appear fuzzy.

    D   They appear brighter.

39 **Which animal is well adapted to a scarcity of sunlight and low temperatures?** S4L1c

    A   polar bear   C   pelican

    B   rattlesnake   D   scorpion

40 Hiroko is making pasta for dinner. She needs to boil water. What temperature must she heat the water to make it boil?

S4E3b

   A   212ºF   B   144ºF   C   100ºF   D   32ºF

41 Thomas is playing on his drum set. How can he make a louder sound?

S4P2a

A   by playing a bigger drum
B   by playing a smaller drum
C   by hitting the drums harder
D   by hitting the drums softer

42 What weather change would NOT occur shortly after a warm front passes over your neighborhood?

S4E4b

A   the temperature rises
B   the air becomes more moist
C   the pressure changes
D   the air becomes cooler and drier

43 What type of simple machine makes a weight feel ½ of its actual weight?

S4P3a

A   screw   C   lever
B   pulley   D   inclined plane

44 Water going through the process of condensation is

S4E3a

A   changing from a liquid to a gas.
B   changing from a gas to a liquid.
C   changing from a solid to a liquid.
D   changing from a solid to a gas.

45 **How would an abundance of trees affect the animal pictured here?**   S4L1d

- A Its population would increase.
- B Its population would decrease.
- C Its population would stay the same.
- D Its population would become scarce.

46 **Rashad records the weather three times a day. In the morning, it is 75°F and dry. A warm front passes through at noon. Which of the following statements did he MOST likely make at noon?**   S4E4b

- A 60°F and dry
- B 70°F and wet
- C 75°F and dry
- D 85°F and wet

47 **Carnivores ALWAYS eat what type of food?**   S4L1b

- A consumers
- B producers
- C decomposers
- D plant matter

48 **Mike hits a ball really hard to Shawn. Shawn hits the ball softly back to Mike. Which player hit the ball the farthest?**   S4P3b, c

- A Mike
- B Shawn
- C they both hit the ball the same distance
- D the force of gravity moves the ball the farthest

49 **Select the transparent object below.**   S4P1a

    A   mirror   C   paper

    B   glass   D   cardboard

50 **Why do we experience different seasons?**   S4E2a

    A   The Earth rotates around the Sun.

    B   The Earth revolves around the Sun.

    C   The Earth rotates on its axis.

    D   The Sun revolves around Earth.

51 **When light is transmitted it**   S4P1a

    A   goes straight through a surface.

    B   gets absorbed by a surface.

    C   gets bent as it goes through a surface.

    D   bounces back.

52 **In the water cycle, which process sends water vapor up into**   S4E4a
    the atmosphere?

    A   evaporation

    B   precipitation

    C   condensation

    D   runoff

53 **Heavy rain would help which organism below?**   S4L1c

    A   frog   C   earthworm

    B   snake   D   stork

54 **Both wrens and swallows eat insects. How would a decrease**   S4L1d
    in wrens affect swallows?

    A   Swallows would also decrease.

    B   Swallows would increase.

    C   Swallows would be unaffected.

    D   Swallows would move away.

55 Playing the hand bells is a tradition in many cultures. What main characteristic of the bells shown below will determine the pitch of the sound they produce?

S4P2b

A mass

B weight

C width

D height

56 What is the role of bass fish in a community?

S4L1a

A to eat food

B to eat wastes

C to make food

D to hibernate

57 The Northern Hemisphere is tilted toward the Sun. How many hours of day and night occur here?

S4E2c

A Daylight and nighttime are equal.

B There are more hours of nighttime than daylight.

C There are more hours of daylight than nighttime.

D The hours of daylight and nighttime are unpredictable.

58 The masses of four balls are given below. Assume all four balls are the same size. The same force is applied to each ball. Which ball will move the LEAST?

S4P3b

A ball A-1 kg

B ball B-2 kg

C ball C-3 kg

D ball D-4 kg

59 An owl that lives in the forest is dark brown and spotted. What type of external adaptation does this describe?

S4L2a

A   mimicry

B   camouflage

C   warning coloration

D   hibernation

60 It is summer time. You are going on a hike outside. You want to stay cool. What color shirt should you wear?

S4P1a

A   black   B   white   C   red   D   blue

61 Which animal below would be LEAST impacted by a dry hot summer?

S4L1c

A   duck

B   alligator

C   camel

D   hippopotamus

62 What causes the Moon to be seen in the night sky?

S4P2b

A   The Moon produces its own light.

B   The Moon reflects light from the stars.

C   The Moon reflects light from the Earth.

D   The Moon reflects light from the Sun.

63 Which type of matter transmits sound the slowest?

S4P2a

A   gas

B   liquid

C   solid

D   vacuum

64 A type of fish eats a specific type of cold water algae. Which factor below would be BEST to use in controlling the fish population?

S4L1c

A   increase the amount of sunlight

B   increase the water temperature

C   decrease the water temperature

D   decrease the amount of salt in the water

65 The Solar System contains S4E2d

A   only the Sun and planets.

B   the Sun, along with planets and asteroids.

C   only the planets.

D   the Sun, planets, and all the stars in the night sky.

66 A type of frog in Canada can survive frozen underground all winter. What type of behavioral adaptation does this frog have? S4L2a

A   mimicry

B   protective

C   hibernation

D   migration

67 Which predator below would MOST benefit from an abundance of gazelles? S4L1d

A   shark

B   rattlesnake

C   frog

D   cheetah

68 Water droplets in a cloud melt and refreeze to form layers of ice. What form of precipitation is this? S4E3d

A   sleet

B   snow

C   hail

D   rain

69 The planets revolve around the Sun because S4E2c,d

A   the Sun is the largest object in the solar system.

B   the Sun is the only star in our solar system.

C   the planets are too small to have their own gravitational force.

D   the planets are the same size with the same gravitational force.

70 Look at the picture below. Which force should he apply to the bowling ball to knock down the most pins?

S4P3b, c

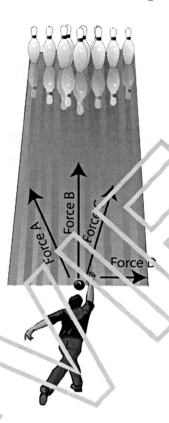

A   Force A

B   Force B

C   Force C

D   Force D

moons 32

N

nimbus cloud 48
North Star 22

O

omnivore 116
opaque 78
orbit 29
organism 111
Orion 22

P

phase change 40
phases of the Moon 32
pitch 89, 90
planet 26, 29
plutoids 30
Polaris 22
population 111
precipitation
   types of 49
predator 116
prey 116
prism 81
probe 26
producer 112, 115
protective adaptation 125
pulley 97

R

rain 49
rain gauge 64
rainbow 82
ramp 94
reflected 77
refracted 76
resistance arm 96
revolution 37
revolve 29, 36
rotate 35
ROY G BIV 82
runoff 54

S

scarcity 132
screw 96
shaft 96
simple machine 93
   types of 93
sleet 49
snow 49
solar system 27, 29
solid 88
sound 87
sound waves 87
sphere 35
stars 19
states of water 39
steam 41
stratus cloud 48
Sun/Earth relationship 36

T

telescope 25
temperature 82
terrestrial planet 30
thermometer 63
theta 76
threads 96
traits 124
translucent 79
transmits 75
transparent 78, 81

U

underpopulated 132
universe 19

V

vacuum 89

W

waning 32
warm front 59
warning coloration 125
water cycle 53
waxing 32
weather 57
weather instrument 63
weather map 58, 60
weather symbol 58
weather vane 65
wedge 95
wheel and axel 97
   examples of 98
white light 81
wind speed 65
work 93